Motor Control Fundamentals Lab Manual

Steve Senty

DELMAR
CENGAGE Learning

Australia • Brazil • Japan • Korea • Mexico • Singapore • Spain • United Kingdom • United States

DELMAR
CENGAGE Learning

Motor Control Fundamentals Lab Manual
Steve Senty

Vice President, Editorial: Dave Garza
Director of Learning Solutions: Sandy Clark
Acquisitions Editor: Stacy Masucci
Managing Editor: Larry Main
Senior Product Manager: John Fisher
Editorial Assistant: Andrea Timpano
Vice President, Marketing: Jennifer Baker
Marketing Director: Deborah Yarnell
Marketing Manager: Katie Hall
Marketing Coordinator: Jillian Borden
Production Director: Wendy Troeger
Production Manager: Mark Bernard
Content Project Manager: Christopher Chien
Production Technology Assistant: Emily Gross
Art Director: David Arsenault
Technology Project Manager: Joe Pliss

Library of Congress Control Number: 2011927125

ISBN-13: 978-0-8400-2463-3

ISBN-10: 0-8400-2463-0

Delmar
5 Maxwell Drive
Clifton Park, NY 12065-2919
USA

Cengage Learning is a leading provider of customized learning solutions with office locations around the globe, including Singapore, the United Kingdom, Australia, Mexico, Brazil, and Japan. Locate your local office at **www.cengage.com/global**

Cengage Learning products are represented in Canada by Nelson Education, Ltd.

To learn more about Delmar, visit **www.cengage.com/delmar**

Purchase any of our products at your local college store or at our preferred online store **www.cengagebrain.com**

Printed in the United States of America
2 3 4 5 6 22 21 20 19 18

Table of Contents

LAB EXERCISE #1

Analog Meter Usage

Name:	Date:	Grade:

Purpose

The purpose of this lab is to review different electrical meter application characteristics, safety practices of meter usage, and the correct methods for making electrical measurements with analog meters commonly used by electricians.

Materials:

Analog multimeter

2–100-kΩ resistors, or greater

2–1-kΩ resistors, or less

1.5-volt dry cell battery

Large capacitor (50 μF or more) rated for 250 volts or more

100-volt DC power supply, or a diode and a 120-volt power source

ANALOG METERS

Basically two different kinds of analog meters are commonly used by electricians: the solenoid tester and the analog multimeter. Solenoid testers are designed to test only for the presence of an AC or DC voltage, whereas an analog multimeter is a test instrument that can be used to test for several electrical parameters, such as AC or DC voltages, resistance or continuity, and small amounts of current in circuits. Because of their multiple functions, the most common problem that occurs when using some types of analog multimeters with multiple lead jacks for different functions or ranges is that the electrician forgets to change the test leads when switching between the different functions of the meter. It is crucial, both to prevent damage and to obtain accurate readings with these types of multimeters, that the test leads be in the jacks corresponding to the test function for which the multimeter is set.

Figure 1-1 Generic analog multimeter

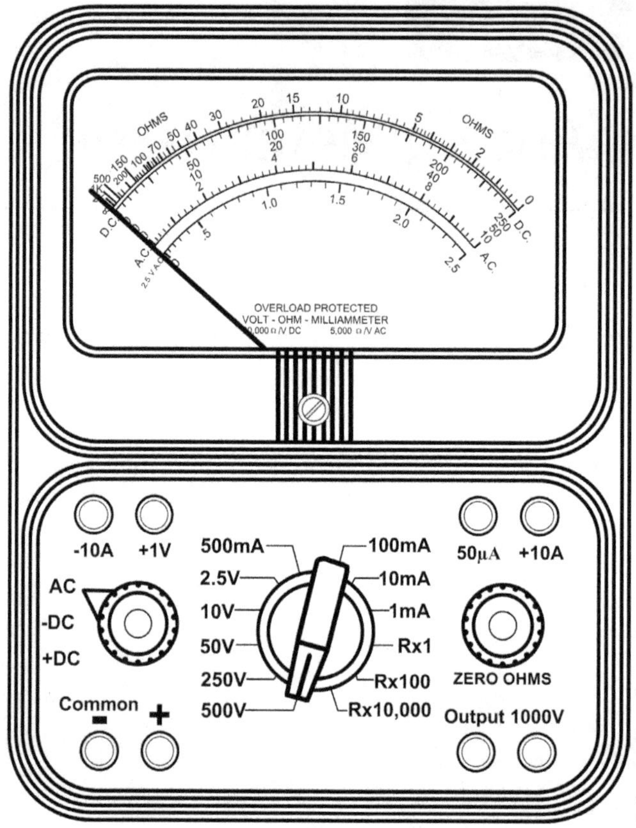

The scales on an analog multimeter are read by means of a needle that moves in front of them, as shown in Figure 1-1; the resistance scale is backward from all the rest of the scales. Many old-timers will tell you that they kept their analog multimeters after they acquired a digital multimeter because the old analog multimeters can bounce around the back of a truck year after year for decades, and they still work when they are needed. Also, analog multimeters can still make a voltage measurement even if the battery is dead–something a digital multimeter cannot do.

Some electricians today dread the thought of having to use an analog multimeter, because they have to set the correct mode (voltage, current, or resistance), the correct range (10 V, 50 V, 250 V, etc.), connect the test leads in the correct jacks, connect the test leads in the circuit for the correct polarity, and then the reading must be interpreted to determine the correct value. Analog multimeters are still being manufactured and sold today, so it is necessary for every electrician to have a working understanding of them in case they are encountered in the field.

As far as input impedance is concerned, the most common analog multimeters are medium-input impedance instruments, and the input impedance changes with the range setting of the meter. Analog multimeters cannot achieve the very high input impedances of digital multimeters without specialized circuitry, because current from the circuit under test must flow through D'Arsonval movement to move the needle, and if the impedance were too high, the movement would not have sufficient current to operate. Analog multi-meters have high enough input impedance, however, that they will render accurate readings in electrical power circuits, and they may take readings indefinitely in an energized circuit without being damaged.

HOW TO INTERPRET ANALOG MULTIMETER READINGS

The first thing to remember about analog meters is that there is not a separate scale on the meter face for every scale selection of the meter selector switch. Notice that the range selector switch in Figure 1-1 shows 12 different positions, but there are only four numbered scales on the meter face. This may seem confusing at first, but each scale on the meter face performs multiple functions, and that is why it is necessary to interpret most readings made with an analog meter. The good news is that the only math involved in interpreting the readings is moving the decimal point of the answer. For instance, the 250 scale of the meter face shown in Figure 1-2 will be used for the 2.5-V and 250-V settings; the 50 scale will be used for the 500-mA, 50-V, and 500-V settings; the 10 scale will be used for the 1-mA, 10-mA, 100-mA, and 10-V scales; and the Ohms resistance scale on the top will be used for the Rx1, Rx100, and Rx10,000 settings.

For example, if the meter range selector switch is set to the 2.5-volt range, and the meter needle moves to the A pointer point shown in Figure 1-2, the reading would be interpreted as follows. First, the correct meter face scale must be chosen. Look at the full scale readings and ask which number—250, 50, or 10—can be changed to 2.5 by moving the decimal point; the answer, of course, is that 250 may be changed to 2.5 by moving the decimal point two places to the left. The reading, then, will be made on the 250 scale of the meter face. Pointer A is indicating 75 on the 250 scale, so the decimal place is moved two places to the left on 75 to become 0.75, or 0.75 volts. This meter does not have a 2,500 scale, but if it did, pointer A would still indicate a value of 75 on the 250 scale. To change 250 into 2,500, the decimal point would have to move one place to the right. When the decimal point is moved one place to the right of 75, it renders a reading of 750, or 750 volts.

An example of the resistance scale is only slightly different. If the meter range selector switch is set to the Rx100 range, and the meter needle moves to the same pointer position A, and noting that the resistance scale reads backwards of the other scales, the indicated value is approximately 26. The answer, however, is 26 times 100, or 2,600 Ω, or 2.6 kΩ. Even though the math is simple, it is even simpler to take the number of zeros in the range setting—there are two zeros in 100 (the Rx100 setting)—and add them to the back of the reading; 26 with two more zeros equals 2,600. If the same reading were made with the meter range selector switch set to the Rx10,000 setting, the answer would be 26 with the four zeros of the Rx10,000 setting added, or 260,000 Ω, or 260 kΩ.

Figure 1-2 Meter face for readings interpretation

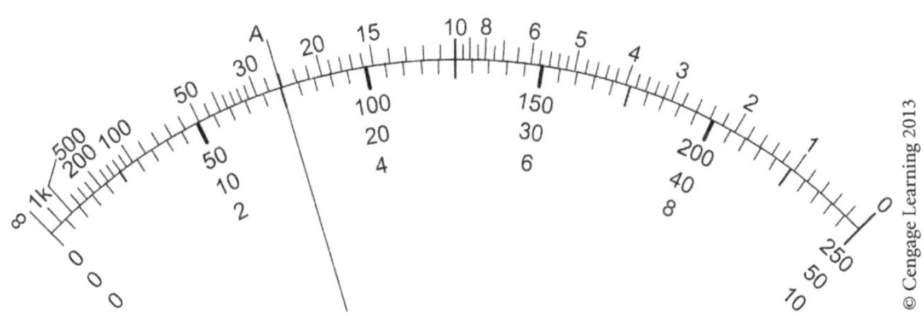

Use these meter range selector switch settings and pointer scale reading methods to interpret the values in Table 1-1 for practice.

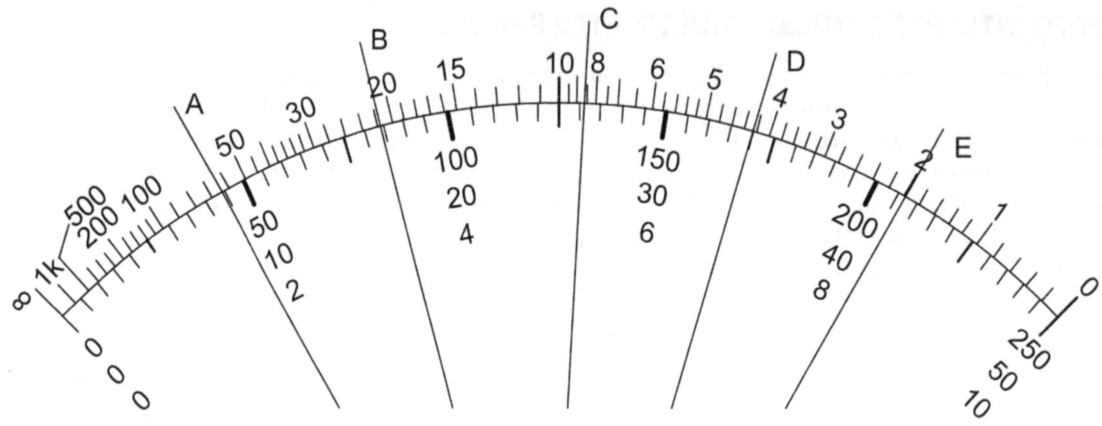

	Point	Range	Reading		Point	Range	Reading		Point	Range	Reading
1	A	10A		18	C	25V		35	E	0.25V	
2	B	50A		19	D	RX10		36	A	0.1A	
3	C	250V		20	E	500A		37	B	250A	
4	D	1V		21	A	2.5V		38	C	RX100	
5	E	5V		22	B	RX1		39	D	500V	
6	A	RX10		23	C	0.5A		40	E	RX100	
7	B	1V		24	D	2.5V		41	A	2500A	
8	C	RX1		25	E	0.25A		42	B	0.5A	
9	D	2.5V		26	A	50A		43	C	2.5A	
10	E	100A		27	B	5OV		44	D	0.1A	
11	A	500V		28	C	50V		45	E	RX10	
12	B	RX10		29	D	RX1		46	A	RX1	
13	C	10A		30	E	RX10		47	B	250A	
14	D	RX100		31	A	2500V		48	C	0.25V	
15	E	250V		32	B	0.1A		49	D	25A	
16	A	RX1		33	C	1A		50	E	RX1	
17	B	5A		34	D	2.5A		51	A	1A	

Table 1-1 Analog meter reading practice

Besides the differences in the way analog multimeters and solenoid testers display their readings, the biggest difference between these types of meters, for the electrician, is input impedance. Input impedance is the amount of load the meter adds to the circuit being tested, which in some cases (not common to electricians) can distort the measured value. A test meter with higher input impedance is not always best for electricians working with transformers or generators, where there is a high level of magnetic flux in the air, because the high input impedance of the meter's circuitry may not drain off an induced or residual electrical charge remaining in the circuit under test. As a rule of thumb, a higher input impedance is better for testing in an energized circuit, because it will change the circuit impedance the least; a lower input impedance is better for verifying that a circuit is de-energized, because it will load the circuit sufficiently to drain off any induced or residual electrical charges that may be present.

Solenoid testers, as shown in Figure 1-3, have the spring-loaded plunger in the middle to indicate the measured voltage value, and they rattle when shaken. When the test leads are placed across a potential difference, current will be pushed through the solenoid coil in the tester, and the magnetic field created by the current will pull the plunger into the core, against the spring tension holding the plunger out of the core. The higher potential difference across the leads, the greater the current flow through the coil, and the further the plunger will be pulled into the core.

Some electricians do not consider the solenoid tester to be a serious meter, because it is not possible to differentiate between voltages such as 208 V and 240 V, and solenoid testers are not useful for reading voltage values under 120 V. In reality, solenoid testers are an indispensable tool for electricians working with transformers and generators, where there is a high level of magnetic flux in the air. Solenoid testers have the lowest input impedance of all the meters, and they load the circuit under test to draw off any induced or residual electrical charges, thus giving a true indication of the energized or de-energized state of the circuit. Solenoid testers have such low impedance, and load the circuit under test so much, that they are rated for the length of time they can be connected in an energized circuit—usually about 15 seconds.

Figure 1-3 Solenoid tester

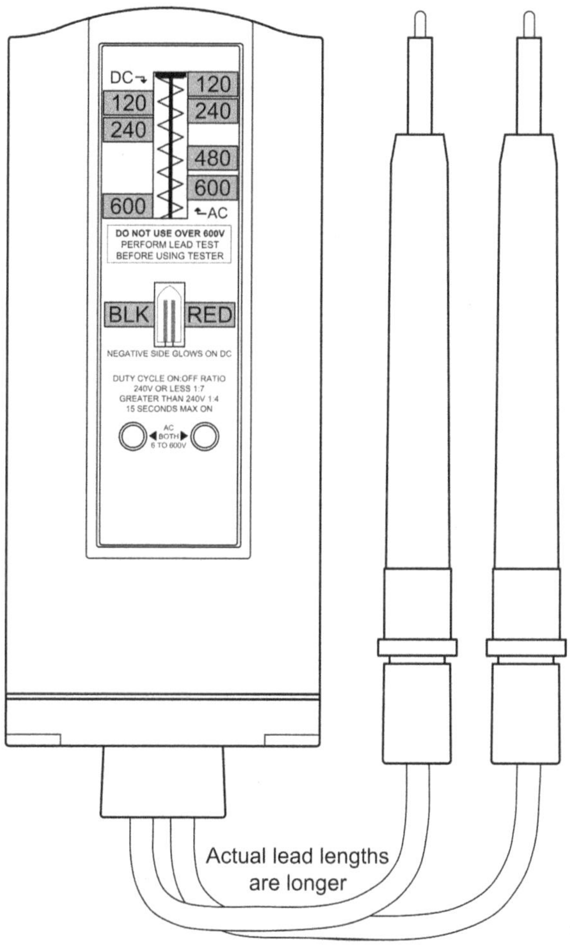

METER INPUT IMPEDANCE LAB ON AN ENERGIZED CIRCUIT

The best way to demonstrate the input impedance effect of a meter on an energized circuit being tested is to exaggerate the effect to make the point clear. The following experiment will show that inserting a low-input-impedance meter into an energized electrical circuit may change the circuit enough to affect the meter readings.

Step 1

Connect the circuit shown in Figure 1-4. With two resistors of equal value connected in series, the supply voltage will divide itself equally between the two. When taking a voltage reading across either of the two resistors connected in series, the technician would expect to read 0.75 volts.

Figure 1-4 Two-100-k resistor circuit

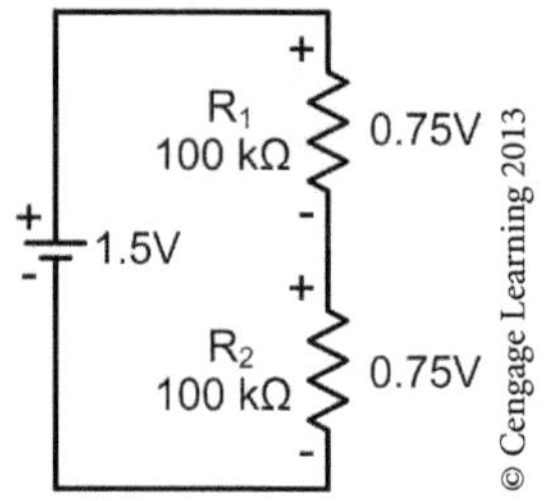

Step 2

Connect the voltmeter across R_1 of the circuit, as shown in Figure 1-5. This lab is going to assume the analog meter you use is 20 kΩ per volt. With the range selector switch set to 2.5 volts, the input impedance of the meter at this setting would be 50 kΩ (20 kΩ × 2.5 volts). When the voltmeter is connected across R_1, it creates a parallel path between R_1 and the voltmeter input resistance. The circuit will change dramatically, because when resistors are connected in parallel the total resistance of the parallel path will be smaller than the smallest resistance—in this case, less than 50 kΩ.

Figure 1-5 Two-100 k resistor circuit with voltmeter

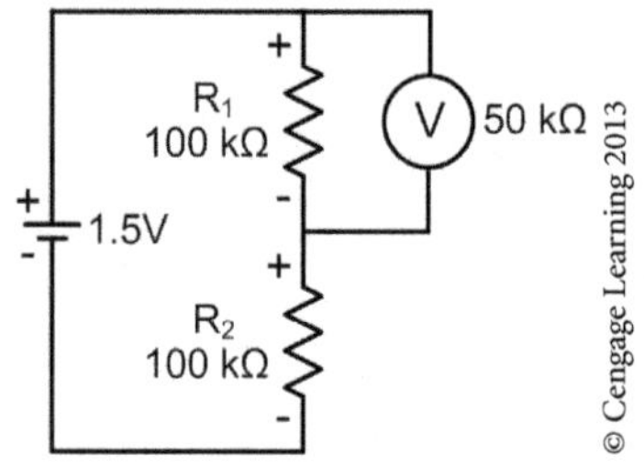

When the 50-kΩ input impedance voltmeter is connected in parallel with R_1, a 100-kΩ resistor, the equivalent resistance of the two will equal:

$$R_E = \frac{1}{\dfrac{1}{100\ \text{k}\Omega} + \dfrac{1}{50\ \text{k}\Omega}} = 33\ \text{k}\Omega$$

The new equivalent circuit will look like the circuit shown in Figure 1-6. Rather than reading 0.75 volts, the analog meter will actually read 0.375 volts. The circuit has been changed by the meter so that an accurate measurement cannot be made.

$$IT = 1.5\ \text{V} / 133\ \text{k}\Omega = 11.25\ \mu\text{A}$$

$$E_{ER} = 11.25\ \mu\text{A} \times 33.3\ \text{k}\Omega = 3.75\ \text{V}$$

Figure 1-6 Two-100-k resistor equivalent circuit

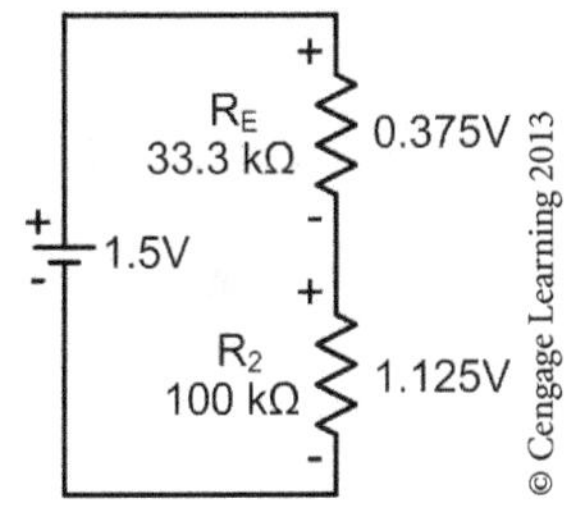

Actual meter reading: ________________ volts.

Step 3

To assure that a meter is going to provide sufficiently accurate measurements, the rule of thumb is that the input impedance of the meter should be at least 10 times greater than the resistance of the component being measured. To test this rule, connect the circuit shown in Figure 1-7, using 1-kΩ resistors in place of the 100-kΩ resistors. Notice that because the two resistors are of equal value, the technician would still expect to read half of the supply voltage, 0.75 volts, across each.

Figure 1-7 Two-1-k resistor circuit

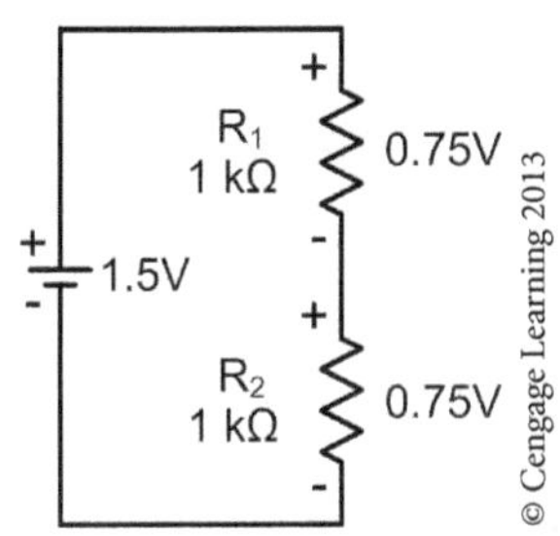

Step 4

Connect the same 50-kΩ input impedance voltmeter used in the last steps into this circuit across R_1, as shown in Figure 1-8, and the circuit will again be changed. The 50-kΩ input impedance of the meter connected in parallel with the 1-kΩ R_1 resistor will create an equivalent resistance of the pair.

$$R_E = \frac{1}{\frac{1}{1\ \text{k}\Omega} + \frac{1}{50\ \text{k}\Omega}} = 980\ \Omega$$

Figure 1-8 Two-1-k resistor circuit with voltmeter

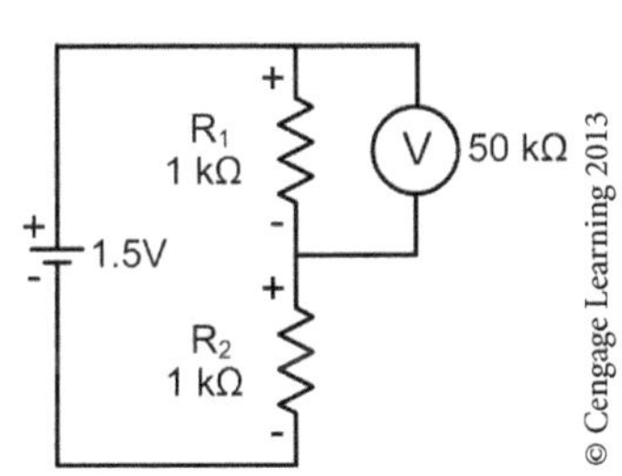

The new equivalent circuit will look like the circuit shown in Figure 1-9. Rather than reading 0.75 volts, the analog meter will actually read 0.743 volts. The circuit has been changed by the meter, but the original circuit is not altered to the point that the meter reading is overly compromised. The new calculation would be as follows:

$$I_T = 1.5\text{ V} / 1{,}980\ \Omega = 758\ \mu\text{A}$$

$$E_{ER} = 758\ \mu\text{A} \times 980\ \Omega = 0.743\text{ V}$$

Actual meter reading: ______________ volts.

Figure 1-9 Two-1-k resistor equivalent circuit

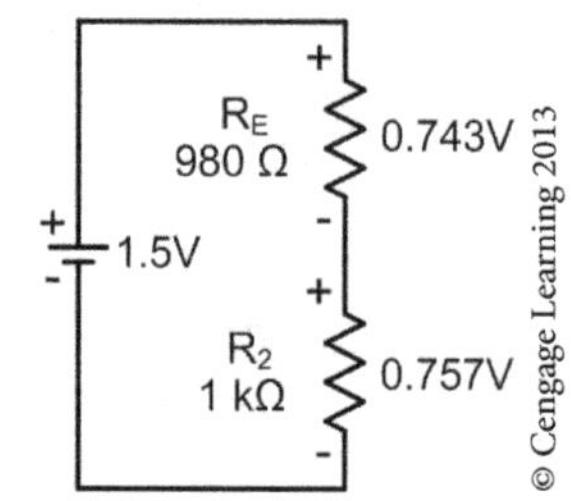

METER INPUT IMPEDANCE LAB ON A DE-ENERGIZED CIRCUIT

Meter input impedance plays an equally important role for electricians when determining the de-energized state of an electrical circuit as it does in making valid voltage readings in energized electrical circuits, but for different reasons. Usually when an electrical circuit is de-energized, no residual static charges from charged capacitors or induced voltages remain in the circuit, and any test meter will indicate that the circuit is de-energized. There are times, however, where residual static charges do remain, and the circuit will appear to still be energized when using a test meter with very high input impedance. When verifying that a circuit is de-energized, a low-input-impedance meter is best, because it will drain off any remaining static charges, and show the circuit as de-energized. The following experiment will demonstrate that using a meter with a very high input impedance to test a de-energized electrical circuit will create problems in verifying the de-energized state.

Step 5

Connect the circuit shown in Figure 1-10, and charge the capacitor to approximately 100 VDC. The actual voltage is unimportant, so if the charging is done with diodes and 120 VAC, the DC charge voltage will be higher.

Figure 1-10 Capacitor charging circuit

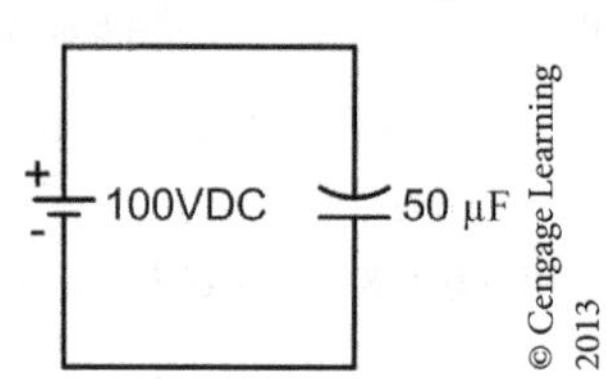

Set the analog multimeter to the highest voltage range setting on the meter; this experiment is going to assume that there is a 250-V setting, and the meter is 20 kΩ per volt input impedance. The input impedance of the analog meter at this setting is 5 MΩ (20 kΩ × 250 V). In this case the circuit is de-energized (removed from the power source), but unless there are other provisions in the circuit, such as a bleeder resistor across the capacitor terminals, the test meter must drain off the residual charge to show the circuit as de-energized.

If a residual charge does remain in an electrical circuit after it is de-energized, it will be dissipated by the resistance-capacitance (RC) time constant created by the input impedance of the meter and the capacitance of the circuit. One RC time constant, measured in seconds, is calculated by multiplying the resistance times the capacitance expressed in Ohm's and farads; in the case of this experiment, 5 MΩ × 50 μF (5,000,000 Ω × 0.00005 F). In this case, 1 time constant is equal to 250 seconds, and it takes 5 time constants to discharge the circuit.

Step 6

To prove this, connect the analog multimeter set to the highest voltage setting across the charged capacitor, as shown in Figure 1-11. Does the meter verify that the circuit is de-energized, or does the circuit appear to still be energized?

Figure 1-11 Capacitor and voltmeter circuit

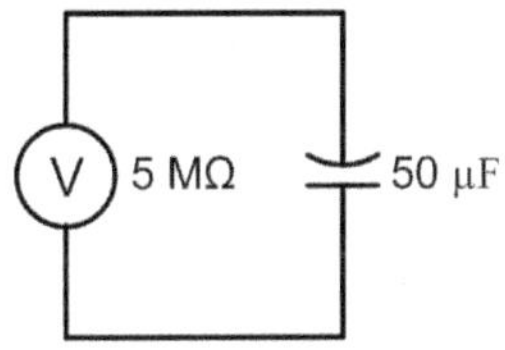

Step 7

Now connect a solenoid type of tester across the capacitor and note the results. Does the capacitor circuit appear to be energized or de-energized?

The reason the solenoid tester shows that the circuit is de-energized is because its input impedance is very low, usually about 5 kΩ. At 5 kΩ the circuit's RC time constant would be 0.25 seconds (5,000 Ω × 0.00005 F). The lower-input-impedance meter, then, will drain off any remaining static charges that can sometimes exist on a de-energized circuit, and indicate the true state of the circuit.

ADVANTAGES AND DISADVANTAGES OF ANALOG METERS

Advantages:

1. The movement of the analog multimeter needle allows easier monitoring of value changes in the test circuit, because needle movement changes are slow and smooth.
2. In most cases analog meters load the test circuit more, and are less susceptible to induced or residual voltage charges.

Disadvantages:

1. The analog multimeter must be manually set up for mode (voltage, current, or resistance) and range (10 V, 50 V, 250 V, etc.).
2. Because there are more terminal options for plugging in test leads, there are more meter setup configurations.
3. Some scales are common to more than one range setting or function, and are thus easily confused.
4. The Ohm's scale is backward from the others, sometimes causing reading errors.
5. The Ohm's scale must be re-zeroed each time the resistance range is changed.
6. Polarity must be observed when connecting the analog multimeter.
7. Analog multimeter self-protection is normally achieved through an internal fuse, which is inconvenient and may leave technicians stranded if they accidently blow the fuse in the field and do not have a replacement.

ELECTRICAL SAFETY FOR ANALOG TEST INSTRUMENTS

Analog multimeters are commonly connected into, and therefore become part of, energized electrical circuits. The following list contains safety precautions specific to analog multimeter use:

1. Inspect the test leads for damaged insulation or exposed metal. Replace the test leads if any damage is found.
2. Make certain the test leads are in the proper jacks, and the meter is in the proper mode, for the measurement you are making.
3. Set the meter to measure resistance, and touch the test leads together to verify lead continuity. The reading should be very close to zero resistance.
4. Set the meter to a voltage setting higher than a known voltage source, and measure the known voltage source to verify the operation of the meter.
5. When using the probes, keep your fingers behind the finger guards on the probes.

QUESTIONS

1. Circle each of the meter range selector switch settings that the meter face 10 scale would be used for on an analog meter.

   0.25 0.5 1 2.5 5 10 25 50 100 250 500 1000

2. Circle each of the meter range selector switch settings that the meter face 50 scale would be used for on an analog meter.

   0.25 0.5 1 2.5 5 10 25 50 100 250 500 1000

3. Circle each of the meter range selector switch settings that the meter face 250 scale would be used for on an analog meter.

   0.25 0.5 1 2.5 5 10 25 50 100 250 500 1000

4. As a rule of thumb, which is better for making measurements in an energized circuit, a high-input-impedance meter or a low-input-impedance meter?
5. As a rule of thumb, which is better for verifying that a circuit is de-energized, a high-input-impedance meter or a low-input-impedance meter?
6. Go back to Step 2; if the meter is moved from R_1 to R_2, what voltage will the meter read?
7. To assure accurate readings, what is the minimum number of times greater the input impedance of a meter has to be than the resistance of the component being read?
8. What is one major advantage of an analog meter, mentioned in the text, that a digital meter cannot duplicate?

LAB EXERCISE #2

Digital Meter Usage

Name:	Date:	Grade:

Purpose

The purpose of this lab is to review different electrical meter application characteristics, safety practices of meter usage, and the correct methods for taking electrical measurements with digital meters commonly used by electricians.

Materials:

Digital clamp meter

3–120-V lamps in sockets: 40 W, 60 W, and 100 W (or any other three different sizes)

DIGITAL MULTIMETERS

Most electrical test instruments that construction electricians will purchase or encounter on a jobsite today are the digital type. Digital multimeters are often preferred over analog multimeters because they offer plug-and-play self-setup capabilities such as auto-range setting, auto-mode setting, internal self-protection, and a direct reading display. Although bench models have more sophisticated functions and greater accuracy, usually a hand held digital multimeter will have sufficient functions and accuracy for the general electrical measurements performed by construction electricians in the field.

Basically two different kinds of digital meters are commonly used by electricians: multimeters and clamp meters. A digital multimeter, as shown in Figure 2-1, is a test instrument that is able to test for several electrical parameters, such as AC or DC voltages, resistance or continuity, and small amounts of current in circuits. Even though the digital multimeter is common among construction electricians for its many capabilities, without the addition of an auxiliary current clamp the multimeter is limited to reading relatively small levels of current, and the circuit must be broken to insert the meter.

Figure 2-1 Digital multimeter

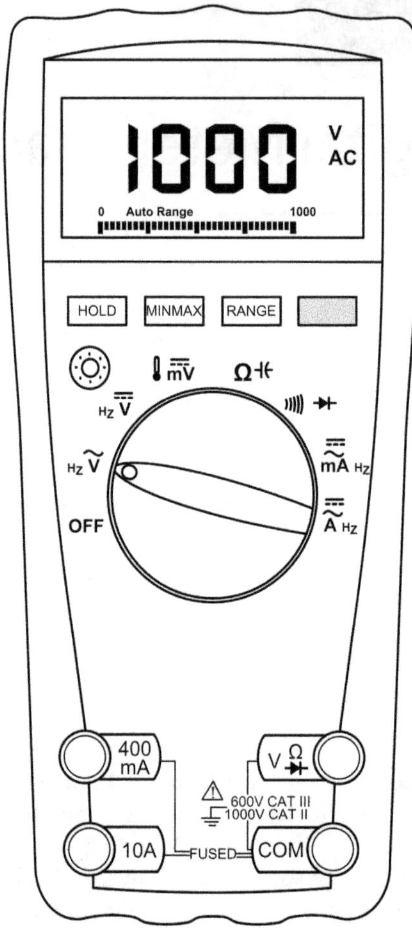

A more popular type of digital meter used by construction electricians is the clamp meter, shown in Figure 2-2. There are two common types of clamp meters. The more common clamp meter on the left of Figure 2-2 has two jacks that allow test leads to be plugged into the meter, which expand the measuring capabilities of the clamp meter to voltage, continuity, and resistance readings, among others. The clamp meter on the right of Figure 2-2 does not have jacks for test leads, and that meter is limited in measuring capabilities to mainly current and frequency.

The clamp meter with test leads is designed to meet the specific needs of construction electricians in the field, in place of a multimeter. The clamp meter has a relatively low input impedance for voltage readings to drain off any residual static charges that may exist on a de-energized circuit, and it uses a current clamp to extend its current-reading capabilities without breaking the circuit to insert the meter. The clamp meter normally has fewer functions than the multimeter, but they are the functions most used by the construction electrician. Figure 2-3 shows the functions of the clamp meter in Figure 2-2 enlarged, with a description of each.

Figure 2-2 Digital clamp meters

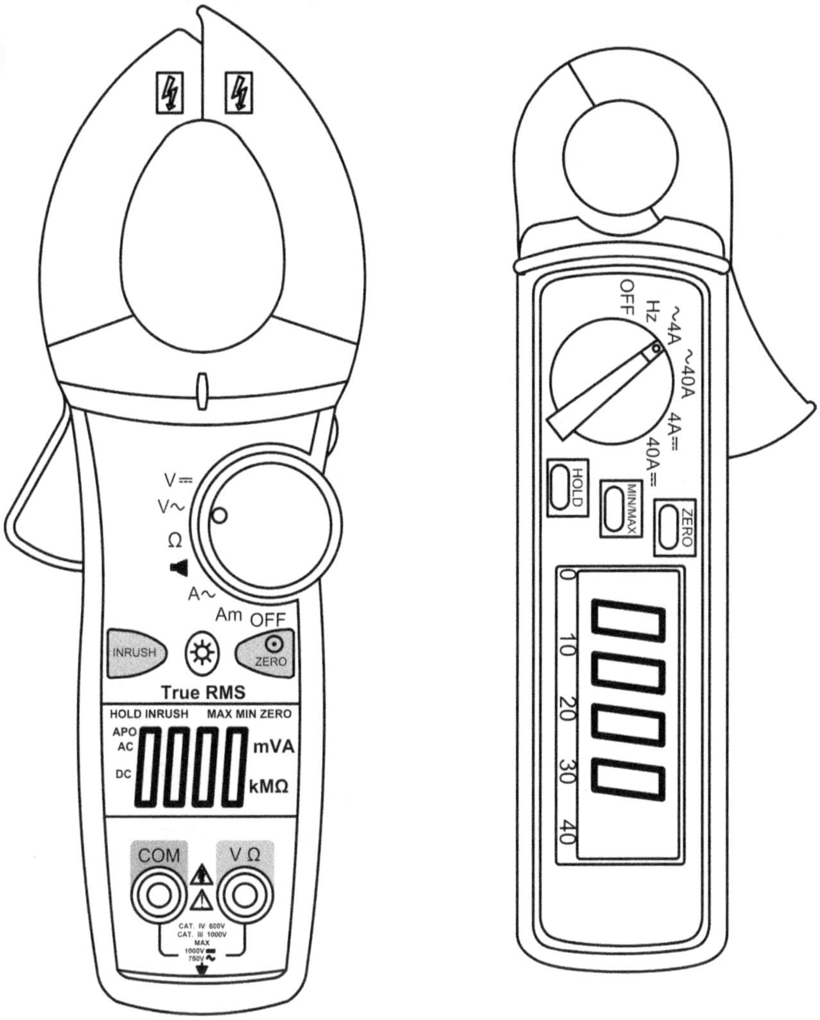

Figure 2-3 Digital clamp meter functions

- V ⎓ DC Volts
- V ∿ AC Volts
- Ω Resistance (ohms)
- Continuity (audible)
- A ∿ AC Amps
- Am Milli-amps

AUTO-RANGING CAPABILITY

The auto-range capability and internal protection of digital meters prevent them from being damaged if they are accidentally connected in the circuit improperly, such as connecting the test leads across a voltage source with the meter set to read resistance, provided the voltage is still within the range of the meter. The biggest advantage of an auto-ranging meter is that it uses the lowest possible range it can to display the signal value being read, which will provide the best resolution, and it determines that range setting without any action on the part of the meter user.

Technicians normally only get into trouble with auto-ranging meters when they try to manually set the range themselves. When the range is set manually, the meter incorrectly assumes that the user knows what he or she is doing, and displays the best reading it can given the manual range setting. With few exceptions, auto-ranging meters provide the best service when no user input is given. When attempting to manually set the range of a digital meter, technicians may find themselves knee-deep in menus of unknown terms, without any idea of what they did to the meter, or even what the meter is now configured to read; in these cases the safest thing to do is to turn the meter off for a couple of seconds and then turn it back on. This will allow the meter to automatically reconfigure itself.

One caution to keep in mind that sometimes causes problems with auto-ranging meters is not paying close enough attention to the meter display. As an example of auto-ranging problems, consider a technician taking voltage readings in the 120-VAC range for a period of time, and the repetition lulls the technician into noticing only the reading value on the display and not the VAC value after it. One of the readings that the technician encounters is 86, but he has not noticed that the meter auto-ranged into the mV range, and it is actually displaying 86 mVAC. The 86 mVAC reading means that the meter is actually reading a de-energized circuit, because digital meters will seldom go down to zero volts when reading a de-energized circuit—it's just a quirk of digital meters. If the technician thought the reading was actually 86 VAC, that would automatically signal a dangerous and unacceptable, possibly open neutral, condition, and the technician could spend hours tracking down a problem that does not exist.

DIGITAL METER DISPLAY RESOLUTION

Display resolution is often specified by the number of digit places the meter can display for a reading. For example, the term *five-digit resolution* means that the meter can display the measured value with five full digits, such as "88888." As a rule of thumb, the more digits of resolution a display has, the more accurate it is. A meter that has a display resolution of one digit could only display values 0 through 9, and only in whole numbers. If a second digit is added, the numbers 10 through 99 will be in whole-number increments, but the values 0 through 9 will gain a 1/10 place accuracy; for example, 4.4 could be displayed. If a third digit is added, the displayed numbers will pick up one more decimal place of accuracy; for example, the previous example of 4.4 could now be shown as 4.36. If a digital meter states that it has a half digit, the most significant digit can display either a zero or a one. There are also digital meters that have a three-quarters digit, but that definition varies by meter manufacturer.

CATEGORY RATINGS

Digital meters also have a category rating, which specifies the maximum transient voltage spikes and overcurrent fault level protection the meter is designed to withstand. The reasoning behind category ratings is that at the facility electrical service entrance, the electrical system's impedance is the lowest, and therefore transient voltage spikes and fault currents would be at their maximum. Further into the facility's electrical system, after going through hundreds of feet of feeders and branch circuits of smaller and smaller wire, the system impedance will significantly attenuate the transient voltage spikes and limit the available fault current. Depending on what part of the electrical system an electrician is working on, a higher protection-withstand rating may be needed to work safely, so the category rating was developed.

There are four categories:

Category I: used where current levels are low, such as within an appliance or machine.

Category II: used on branch circuits.

Category III: used on permanently installed loads, such as distribution panels and motors.

Category IV: used in locations where transient voltage spikes and fault currents would be at their maximum, such as service entrance main panels.

Most electrical test meters designed for construction electricians are category IV up to 600 volts, and may be used safely anywhere in the facility's electrical system.

TAKING A CURRENT READING WITH A CLAMP METER

The purpose of this lab is to demonstrate how to use the clamp meter to take the current measurement of a single component in a circuit with more than one component.

Explanation

As shown in Figures 2-4 and 2-5, clamp the jaws of the meter around all the wires on either terminal of the load to be measured, regardless of how many wires are on each terminal. In both Figures 2-4 and 2-5, only the current of the 50-Ω resistor is being measured. In Figure 2-4 it is easy to see that the current of the 75-Ω resistor is not being measured, because there are two leads on the right side of the 75-Ω resistor, but only one lead goes through the clamp meter. In Figure 2-5, there is only one lead on the left side of the 75-Ω resistor, and it is going through the clamp meter, but the clamp meter is not measuring the 75-Ω resistor current. The reason is that the clamp meter has two wires passing through it, and the 50-Ω resistor is the only component that both of these wires are connected to on one end.

Figure 2-4 One-lead 50-ohm resistor-only current

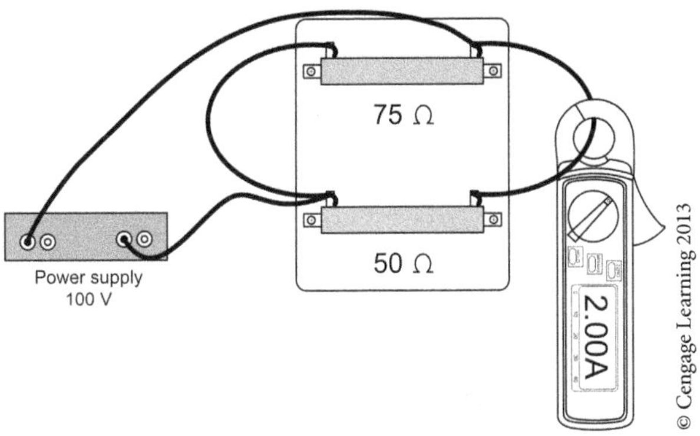

Figure 2-5 Two-lead 50-ohm resistor-only current

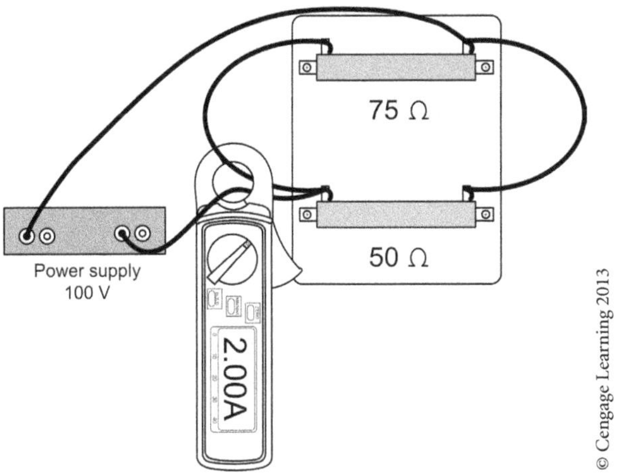

The arrows in the diagram in Figure 2-6 show the current path of the 50-Ω resistor in the circuit, and how the clamp meter in Figure 2-4 would measure it. Notice that the current of the 50-Ω resistor starts at the power supply; touches, but does not pass through, the 75-Ω resistor; goes through the clamp meter and the 50-Ω resistor; and then returns to the power supply.

The arrows in the diagram in Figure 2-7 show the current path of the 50-Ω resistor in the circuit, and how the clamp meter in Figure 2-5 would measure it. Again, the current of the 50-Ω resistor starts at the power supply; touches, but does not pass through, the 75-Ω resistor; goes through the 50-Ω resistor and then the clamp meter; and then returns to the power supply.

Figure 2-6 One-lead 50-ohm resistor-only current explanation

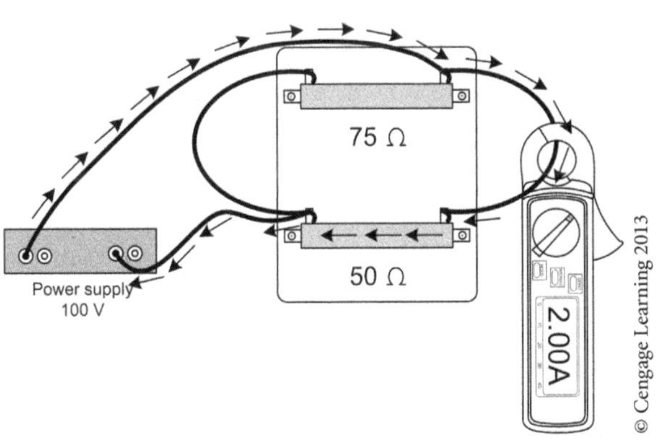

Figure 2-7 Two-lead 50-ohm resistor-only current, 50-ohm current explanation

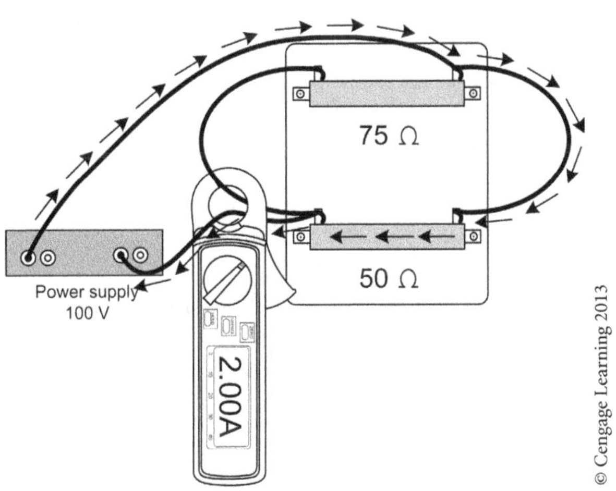

It is apparent that the current of the 75-Ω resistor passes through the clamp meter when the current of the 50-Ω resistor is measured on the left side, so why isn't the current of the 75-Ω resistor measured? Because as shown in Figure 2-8, when measuring the current of the 50-Ω resistor on the left, as shown in Figure 2-7, the current of the

75-Ω resistor passes through the clamp meter in both directions, and the magnetic fields cancel each other. To measure current, clamp meters actually measure the magnetic field created around the conductor by the current flow within the conductor, so the current being measured can only pass through the clamp meter in one direction. If the same current passes in both directions through the clamp meter, the magnetic fields of the two will cancel each other, and the net will be zero. This is why the clamp meter will always read zero if the clamp is placed around both the hot and neutral of the load being measured.

Figure 2-8 Two-lead 50-ohm resistor-only current, 75-ohm current explanation

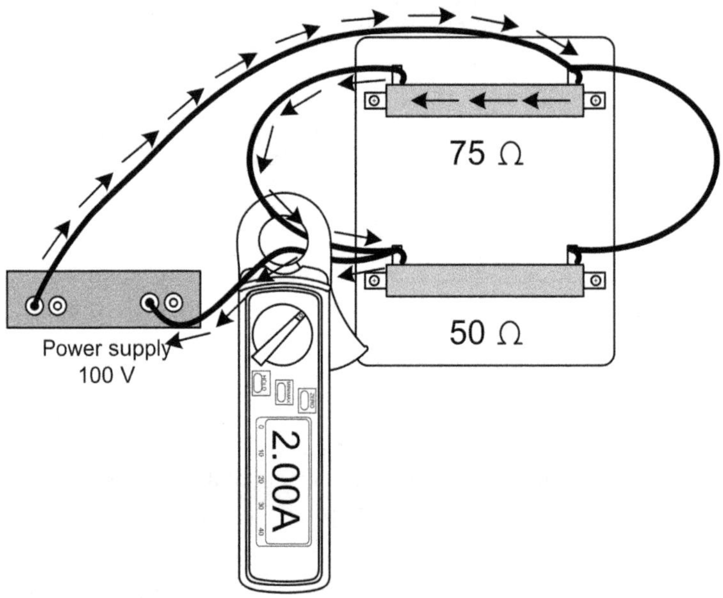

10-LOOP WIRE COILS

Clamp meters that are designed to read hundreds of amperes are not always the best at reading milliamperes. Although the rare occasion does arise, the need to measure milliamperes is certainly not common for construction electricians. It is, however, a common need for students, because it is easier and safer to build lower-power circuits in the lab environment to reinforce electrical theory concepts. The low range of a clamp meter can be extended by simply looping the conductor being measured through the clamp more than once. By doing this, the strength of the magnetic field around the conductor that is created by the small current will be multiplied by the number of times the conductor is looped through the clamp meter. The only thing to remember is that the current value displayed on the clamp meter must be divided by the number of turns you looped through the clamp to get the correct value.

For example, if the wire is looped through the clamp twice, and the meter reads 0.6 amperes, the actual value of the reading is 0.3 amperes (0.6 amperes / 2 loops). In this same example the wire could be looped through the clamp meter 7 times, and the meter would read 2.1 amperes. The technician would then divide 2.1 amperes by 7, and calculate the actual value as 0.3 amperes. Rather than doing the math for each reading, it is easier to purchase or make a coil with 10 loops as shown in Figure 2-9; the actual reading value is then determined by moving the decimal point. With a 10-loop coil, a reading of 3 amperes would actually be 0.3 amperes, or 300 mA; a reading of 0.6 amperes would actually be 0.06 amperes, or 60 mA; and so on.

Figure 2-9 Clamp meter with 10-loop coil

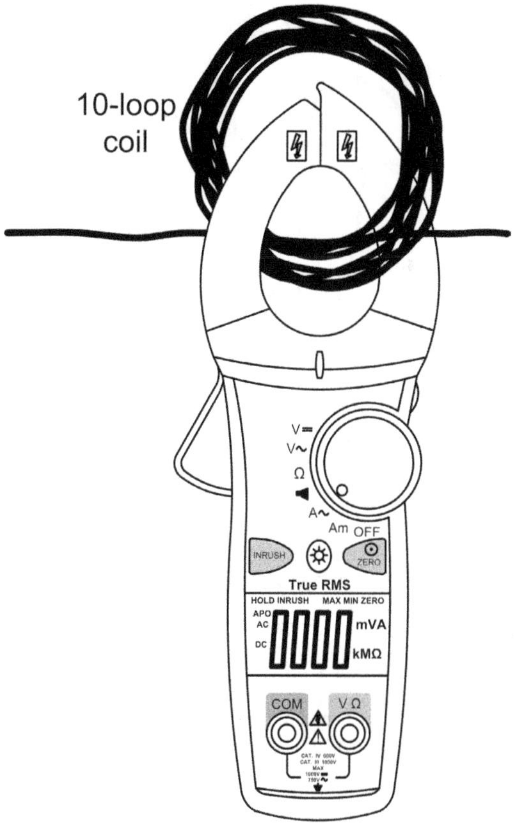

LAB FOR CLAMP METER READINGS

Step 1

Connect the circuit, shown in Figure 2-10, of three unequal-size lamps connected in parallel. The purpose of using unequal-size lamps is to provide three different current levels in the circuit. Be certain to observe the connection differences between the left and right sides of the drawing, as the connection is deliberate. Because voltage is the same in all parts of a parallel circuit, no voltage measurements will be taken for this circuit.

Figure 2-10 Parallel lamp drawing without meter

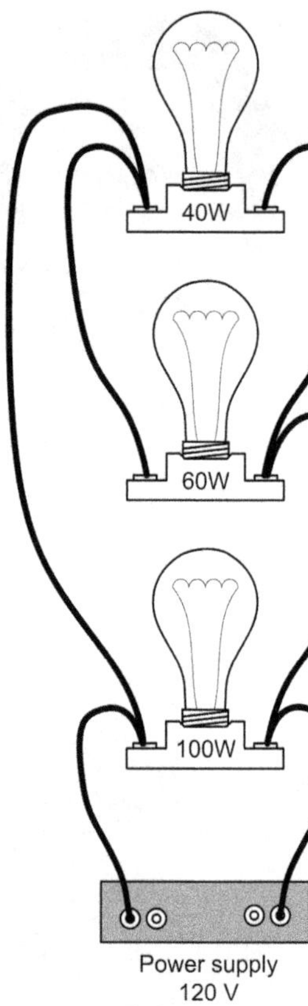

Step 2

Using the formula I = P/E, calculate the current for each of the lamps in Table 2-1.

Calculate Current for Each Lamp		
40 W	60 W	100 W
333mA	500mA	830 mA

Table 2-1 Calculated current for each lamp

Step 3

Use the clamp meter to take the current reading of the 60-watt lamp, first with only the single lead on the left side f the lamp, as shown in Figure 2-11, and then with both of the leads on the right side of the lamp, as shown in igure 2-12. Record the readings in Table 2-2.

Figure 2-11 Parallel lamps—meter on one lead left

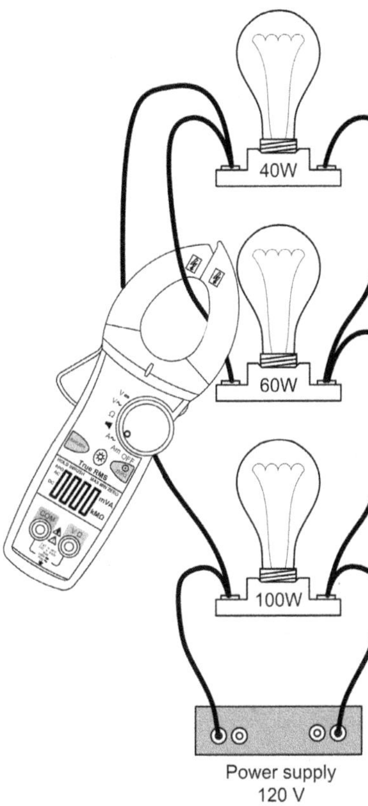

Figure 2-12 Parallel lamps—meter on two leads right

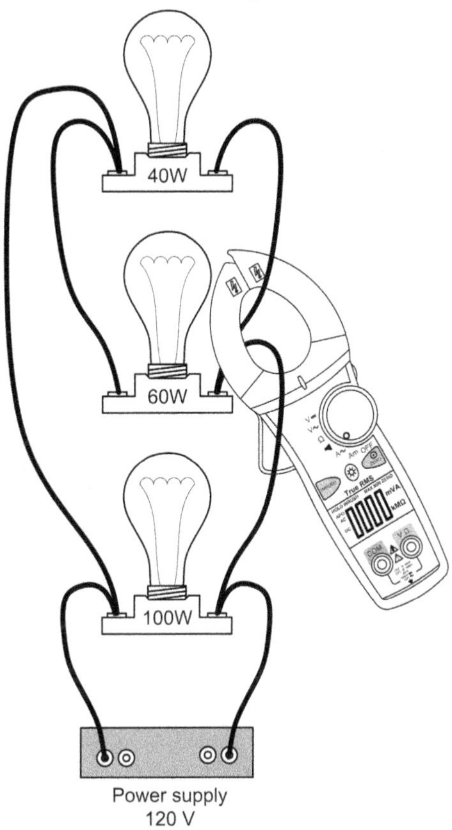

Measurements for 60-W Lamp	
Figure 11 Measurement	Figure 12 Measurement
490 mA	460 mA

Table 2-2 Measured values from Figures 2-11 and 2-12

1. Are the two readings in Table 2-2 the same? NO
2. Do the readings in Table 2-2 agree with the calculated current for the 60-watt lamp in Table 2-1?

   Yes

Step 4

Use the clamp meter to make the current reading of the 60-watt lamp using only one of the two leads on the right side of the lamp, first as shown in Figure 2-13, and then as shown in Figure 2-14. Record the readings in Table 2-3

Figure 2-13 Parallel lamps—meter on top lead only right

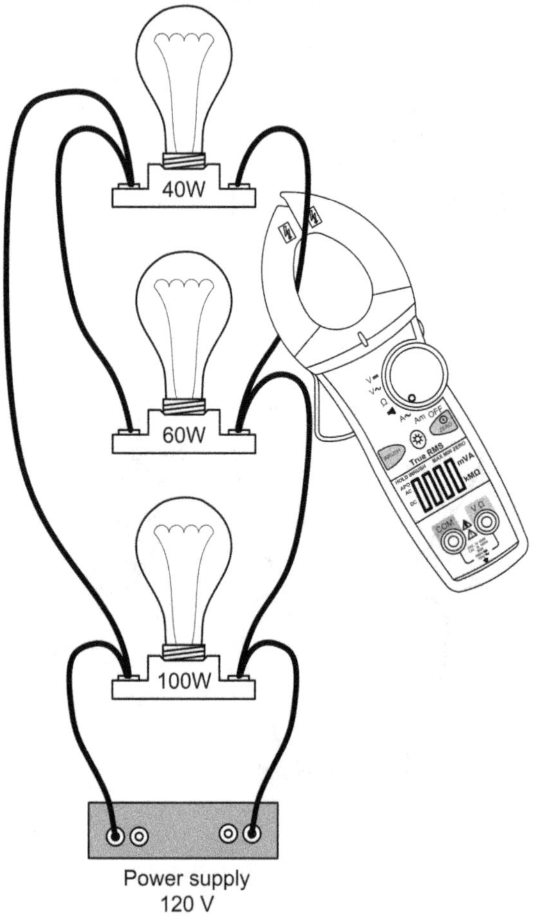

Figure 2-14 Parallel lamps—meter on bottom lead only right

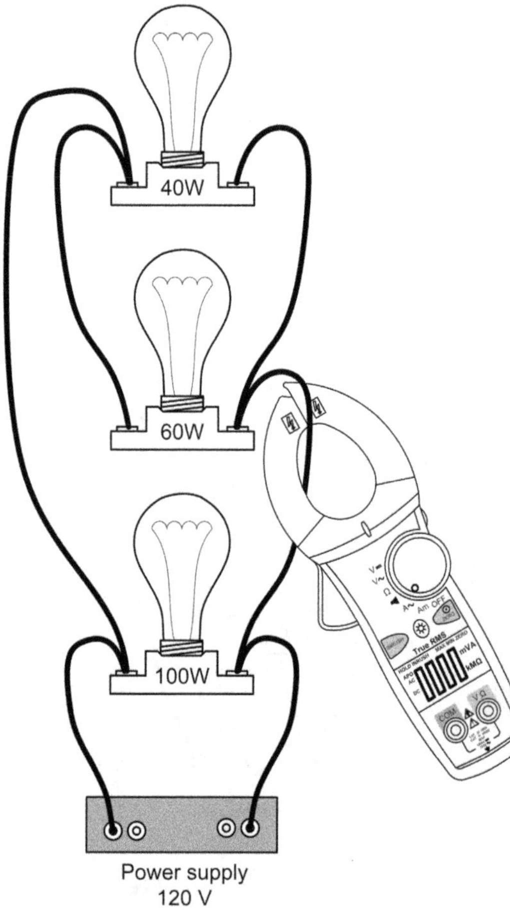

Measurements for 60-W Lamp	
Figure 13 Measurement	Figure 14 Measurement
280 mA	790 mA

Table 2-3 Measured values from Figures 2-13 and 2-14

1. Do either of the clamp meter readings in Table 2-3 agree with the calculated current for the 60-watt lamp in Table 2-1, or the measured values in Table 2-2? yes
2. Does this confirm that all leads on one side of a load must pass through the clamp meter for the correct reading? yes

Step 5

Using the same method with the clamp meter, make the current measurements for Table 2-4. To make the total circuit current measurement, clamp the meter around all of the leads on one side of the power supply.

All Current Measurements			
40 W	60 W	100 W	Total
370 mA	480 mA	790 mA	1.64 A

Table 2-4 All current measurements

1. Do the measured current values in Table 2-4 agree with the calculated current values in Table 2-1 for each of these lamps? yes
2. Did the sum of all of the lamp currents add up to the total circuit current?

yes

Step 6

Figure 2-13 is actually measuring the current of the 40-watt lamp. To make this current measurement, connect a lead between the lamps in the circuit that is long enough to loop through the clamp of the meter 10 times. The current reading for a single wire through the clamp from the 40-watt lamp is already documented in Table 2-4; continue increasing the number of turns through the clamp to complete Table 2-5.

Current Readings of 40-W Lamp from Additional Conductor Loops Through the Meter Clamp								
2 Loops	3 Loops	4 Loops	5 Loops	6 Loops	7 Loops	8 Loops	9 Loops	10 Loops
590 mA	930 mA	1.23 A	1.58 A	1.95 A	2.2 A	2.7 A	2.99 A	3.39 A
Reading/ 2 =	Reading/ 3 =	Reading/ 4 =	Reading/ 5 =	Reading/ 6 =	Reading/ 7 =	Reading/ 8 =	Reading/ 9 =	Reading/ 10 =
295 mA	310 mA	307 mA	316 mA	325 mA	314 mA	337 mA	332 mA	339 mA

Table 2-5 Additional conductor loops through clamp

1. When each of the readings in Table 2-5 was divided by the number of conductor loops through the meter clamp, were the readings consistently equal to the current value read for the 40-watt lamp in Table 2-4? Yes
2. If the readings changed slightly as the number of conductor loops through the clamp increased, would it be reasonable to assume that the readings got more accurate or less accurate?

more accurate

Step 7

Connect the circuit, shown in Figure 2-15, of three unequal-size lamps connected in series. Because current is the same in all parts of a series circuit, no current measurements will be taken for this circuit.

Figure 2-15 Three lamps connected in series

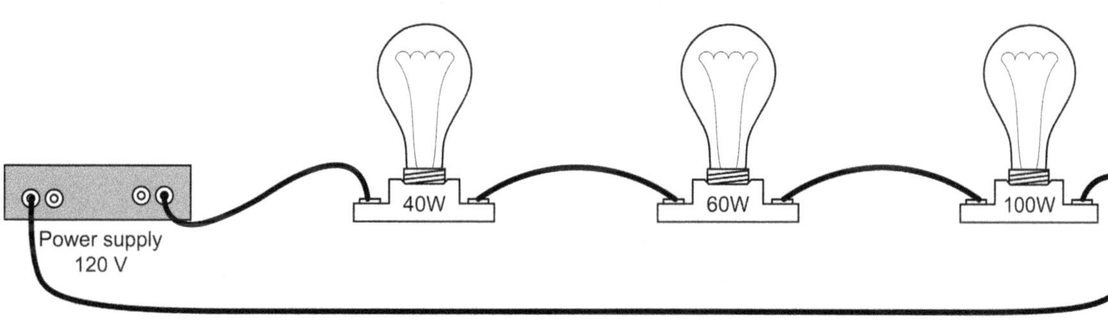

Step 8

The types of clamp meters that construction electricians use tend to have fewer features, and therefore do not require a lot of different test lead jacks. Most clamp-type meters will have only two test lead jacks, which are used for basic features such as voltage, continuity, and resistance, so these meters are almost foolproof. Voltage readings are made by connecting the meter test leads directly across a load or power source, as shown in Figure 2-16. Take the voltage readings necessary to complete Table 2-6.

Figure 2-16 Voltage measurement of 40-W lamp

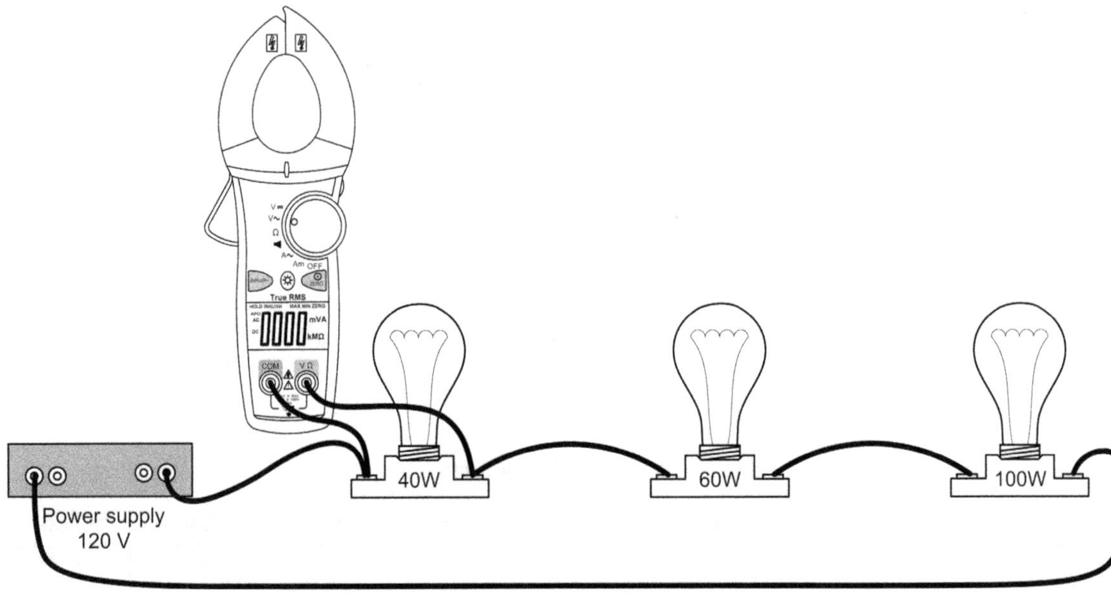

Series Circuit Voltage Measurements			
40 W	60 W	100 W	Supply
73.3 V	34.7 V	12.25 V	120.6 V

Table 2-6 Series circuit voltage measurements

Step 9

In a series circuit, component voltage drop should be proportional to component resistance; the greater the component resistance, the greater the voltage drop. Use the electrical formula $R = E^2/P$ to calculate the resistance of each lamp. It should be noted that this calculation will give the hot resistance of the lamp filament. The lamps in this circuit will not reach their full operating temperature because they will not have their rated voltage applied to them, but we can still identify which lamps have the highest and lowest resistance with this calculation in Table 2-7.

Lamp Filament Resistances		
40 W	60 W	100 W
134.3 Ω	20 Ω	1.5 Ω

Table 2-7 Lamp filament resistances

1. Review Tables 2-6 and 2-7. Did the lamp with the highest resistance have the highest voltage drop?
2. Likewise, did the lamp with the lowest resistance have the lowest voltage drop?

Step 10

A concept of series circuits that is very important to understand when troubleshooting motor control circuits is that the source voltage is read across an open circuit. With the circuit from Figure 2-15 energized, connect the clamp meter to read the voltage across the 40-watt lamp. Carefully unscrew the 40-watt lamp, and document what happens to the voltage across the lamp in Table 2-8 when the lamp is unscrewed. Screw the 40-watt lamp back into the lamp socket, and repeat the process for each of the lamps.

Open Circuit Voltage Across Each Lamp		
40 W	60 W	100 W
120V	120V	120V

Table 2-8 Open circuit voltage across each lamp

1. When each of the lamps was unscrewed from the lamp socket to create an open circuit, did the measured voltage change from the values recorded in Table 2-6 for each of the lamps to the source voltage? yes

ADVANTAGES AND DISADVANTAGES OF DIGITAL METERS

Advantages:

1. Capabilities such as auto-range setting, auto-mode setting, and auto-off to extend battery life
2. Direct read display to eliminate reading interpretation errors.
3. Greater accuracy
4. Self-protection from user mistakes

Disadvantages:

1. The technician must watch the display for automatically changing display values.
2. Digital meters do not always "zero" perfectly, even in a de-energized circuit, so the technician must pay careful attention to the display value.
3. The display digits jump around, and do not show smooth value changes.
4. False readings, which could lead to possible electric shock or personal injury, may occur as the battery becomes very low. Replace the battery as soon as the battery indicator appears.

ELECTRICAL SAFETY SPECIFIC TO CLAMP-ON AMMETERS

1. Never use clamp-on ammeters around electrical circuits of more than 600 VAC (RMS). The clamp-on ammeter does not per se have a maximum voltage for which it can read the current in a circuit; the circuit voltage level is for the dielectric value of the meter case to protect internal conductive parts that the circuit voltage could arc through to.
2. The wire being tested must pass through the jaw, and the jaw must be closed.

QUESTIONS

1. Why are digital meters often preferred by technicians over analog meters? give accurate reading
2. What is meant by the term *display resolution* when discussing digital meters? number of digits
3. What is meant by the term *half digit* when discussing digital meter display resolution? only goes up to a certain number
4. If a technician ever messes up the configuration of an auto-ranging meter, what is the best recovery procedure to follow? restart
5. Which digital meter rating determines the transient voltage spike and overcurrent fault level protection of a meter? cat rating
   a. As the rating number becomes higher, does the protection level increase or decrease? increase
6. Clamp-type ammeters actually determine the current value in a conductor by measuring what? magnetic lines of flux
7. Why would a clamp-on ammeter have a circuit voltage limitation?

LAB EXERCISE #3

Disassemble and Reassemble Motors

Name:	Date:	Grade:

Purpose

The purpose of this lab is to give the learner hands-on experience in disassembling single-phase and three-phase induction motors, examining each individual component part for construction and function, and reassembling the motors.

Materials:

Single-phase motor for disassembly

Three-phase motor for disassembly

MOTOR REPAIR AND REPLACEMENT

A great deal of debate exists among educators in the electrical construction industry concerning the value of using their limited lab time to have students learn about the component parts of electric motors. Most construction electrician programs have stopped teaching motor stator rewinding, rotor growling, and even bearing replacement as part of their normal program curriculum. The reasoning is that electricians simply do not perform repair work on motors enough to justify the time and resource commitment, when that time would be better spent preparing learners for work they are more likely to encounter in the field. It is true that the majority of construction electricians will never perform repair work on a motor; it simply is not cost-effective for the customer.

Mass-produced smaller motors (under 10 horsepower) that go bad are so inexpensive that many customers would rather have the motor replaced with a new motor than send it to a motor shop or have an electrician try to repair it. First of all, unless the problem is something really simple, such as bearings, the labor charges alone would be more than the cost of a new motor, and repair parts may be difficult to find. If the problem is something simple, such as bearings, the customer may be able to save a few dollars repairing the old motor, but the customer ends up with an old motor with new bearings. Replacing smaller electric motors, rather than repairing, probably results in the customer receiving a more efficient, better built, and maintenance-free solution to the motor problem.

There is no magical horsepower size point where it becomes more advantageous to repair an electric motor rather than simply replacing it, but generally it is more cost-effective to repair a larger motor than it is to replace it. This still does not mean that electricians are going to perform much repair work on large motors. If it is a simple repair, such as bearings, electricians may be called on to perform that work, because the repair can often be made in the field without electrically and mechanically removing the motor from its mounting. More often, however, even for bearings, the customer would rather have the motor removed from its mounting and sent to a motor repair shop for repairs. Among the advantages of having a motor repair shop perform the repairs is that the shop will perform other tests on the motor to gauge its overall health, such as insulation tests, that most electricians simply are not trained or prepared to perform.

The intent of disassembling a motor in this lab is not to train technicians in electric motor repair—it is to observe and verify understanding of the parts and operating concepts studied in the textbook. Disassembling and reassembling a motor will give the learner a much better understanding of how all of the different component parts of an electric motor fit and work together, and actually actuating the centrifugal weight assembly on the shaft of a single-phase motor will contribute more to understanding than any number of pictures in the textbook.

DISASSEMBLE A SINGLE-PHASE MOTOR

Step 1

Acquire a single-phase motor to disassemble. Figure 3-1 shows a single-phase capacitor-start motor, which will be used for demonstration purposes here; the motor you use may look much different. Without looking at the motor nameplate, it is possible to determine the type of motor; the capacitor doghouse on the top of the motor indicates it is a capacitor motor, and only single-phase motors have capacitors.

Figure 3-1 Single-phase capacitor-start motor

If the motor has never been apart before, it is important to mark the end bells and motor housing with line-up marks, as shown in Figure 3-2, to make certain the motor is reassembled with the parts in the correct positions in case they are not interchangeable. On smaller motors like this one, the line-up marks are most often made on the bottom of the motor where they will be out of sight when the motor is installed, but on larger motors the marks are often on the top for easy sighting. As shown, one end bell is marked against the motor housing with only one mark, and the other end bell is marked with two marks; this will assure that the correct end bell ends up on the correct end of the motor housing when the motor is reassembled. In this case the line-up marks are made with a permanent marker, but if there is a worry that the marks could accidently be wiped off, they could be made with a scratch awl, or possibly a center punch—a single punch on one end bell and motor housing, and two punches on the second end bell and motor housing. If the motor has been disassembled before, these marks probably already exist, but be certain to take note of them.

Figure 3-2 Line-up marks on bottom of motor

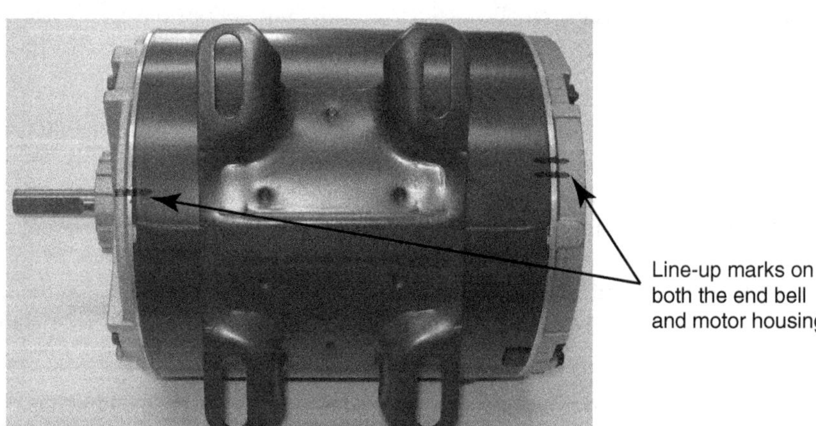

Step 2

Observe what different types of conductor connection methods the motor is designed to use. Set the motor on its mounting feet, and remove the terminal box cover, as shown in Figure 3-3. Notice the grounding screw inside the terminal box; every motor should have a grounding provision inside the motor terminal box so that the grounding conductor does not have to leave the terminal box to ground the motor, where it would not be protected from physical damage. This motor has post and nut connections to land solid wires directly, or crimp lugs on stranded wires. Because this is a relatively small motor, quick-connect terminals are also provided as a connection method.

Figure 3-3 Motor terminal box

1. Did you locate the grounding provision in the motor terminal box of your motor?
   a. What is the grounding provision—screw, lug, etc.?
2. What conductor terminating methods is your motor designed to use?

Step 3

Observe what different types of wiring methods the motor is designed to use. As shown in Figure 3-4, this motor is designed to accept two different sizes of cord. Once the cord is stripped and terminated on the motor terminals, the un-stripped part of the cord with the outer rubber jacket still attached is placed in one of the openings, and replacing the terminal box cover clamps the cord securely. Figure 3-5 shows an electrical fitting hub on the motor that can also be used to access the motor terminal box by removing the rubber grommet covering the hole. This provision is made for wiring methods that use an electrical fitting, such as flexible conduit or a cord connector.

Figure 3-4 Provisions for cord wiring

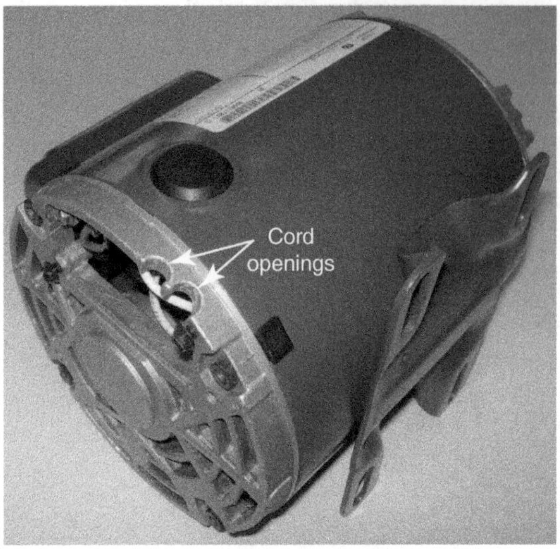

Figure 3-5 Electrical connector fitting hub

1. What different types of wiring methods is your motor designed to use?

Step 4

The first step in disassembling a motor is to identify and remove the fasteners holding the end bells on the motor housing. As shown in Figure 3-6, this particular motor uses thru bolts, which extend from one end bell to the other, to hold the end bells in the motor housing. Remove whatever fasteners are being used to hold the end bells on the motor you are disassembling.

Figure 3-6 Removing the thru bolts

Step 5

Remove the end bells. The end bells fit very tightly inside the motor housing to keep the end bells, and therefore the motor shaft, centered in the motor stator. After the fasteners holding the end bells on the motor housing are removed, the end bells may require a little motivation to break loose. Figure 3-7 shows screwdriver slots provided on the bottom of this motor. Inserting a screwdriver into this slot and turning it will cause the end bell to separate from the motor housing if the fasteners are removed.

Figure 3-7 Screwdriver slot

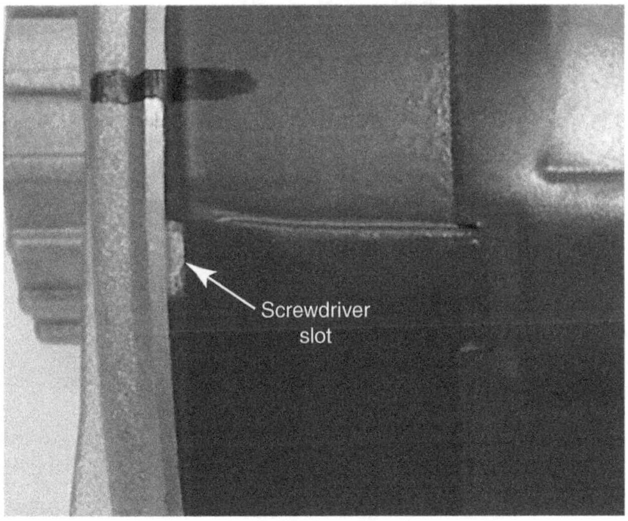

1. What provision for loosening the end bells from the motor housing does your motor have?

Figure 3-8 shows the motor end bells loose from the motor housing, and ready to remove.

Figure 3-8 End bells loose from motor housing

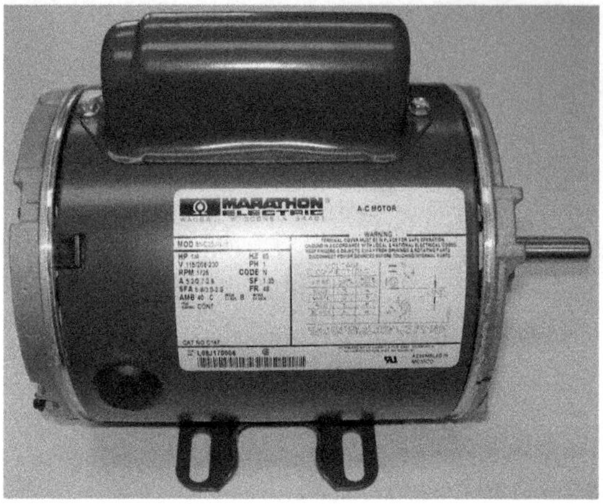

Step 6

Remove the non-drive-end end bell from the motor housing, as shown in Figure 3-9. Looking inside the non-drive-end of the motor with the end bell removed, notice that the connection terminal block and the centrifugal weight assembly are both on this end of the motor. Notice also that on this end of the motor the rotor shaft ends at the back of the motor bearing.

Figure 3-9 Non-drive-end end bell

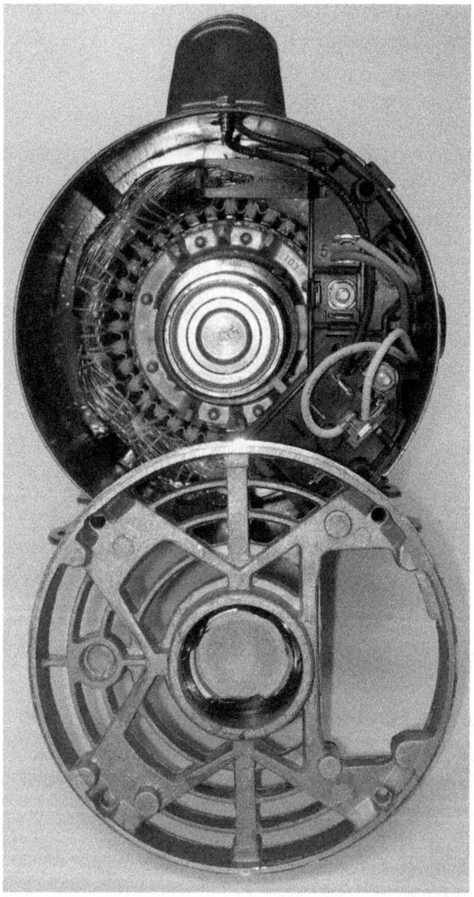

Figure 3-10 shows the non-drive-end end bell alone. Notice that the end bell has a hub that is designed for the rotor shaft bearing to fit into. The hub should hold the bearing snug, and prevent the outer race of the bearing from spinning inside the hub when the shaft turns. If the end bell bearing hub is slightly worn and the outer race of the bearing turns inside it, the hub will wear out very quickly, as this is not a designed wear point.

Another point of note with the non-drive-end end bell is what is inside the bearing hub. On the back of the hub there is a special washer, called a wave spring washer, which is shown in Figure 3-11. Wave spring washers get their name from their curved design, which allows them to provide a spring effect to absorb shock or movement. The wave spring washer is installed in the bearing hub, and pushes against, or pre-loads, the outer race of the bearing, keeping the rotor pushed against the drive-end side of the motor. This ensures that the rotor will not move back and forth inside the motor, and provides smoother and quieter operation.

Figure 3-10 Non-drive-end end bell alone

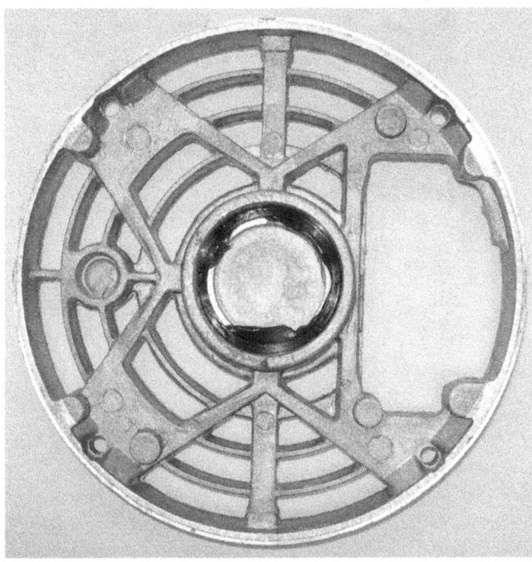

Figure 3-11 Wave spring washer

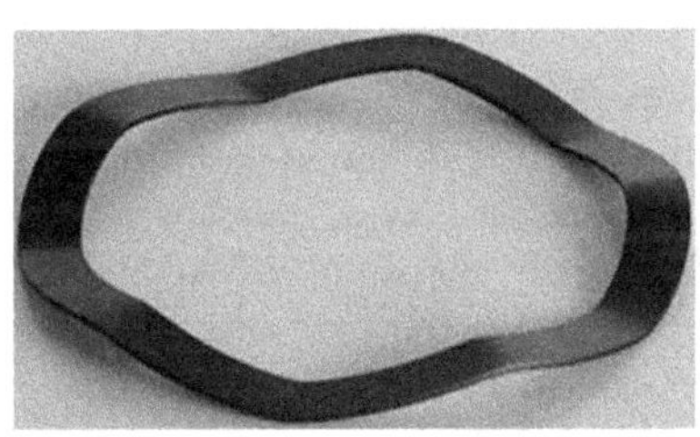

1. Is there a wave spring washer inside the bearing hub of the non-drive-end end bell of the motor you have?
2. Remove the wave spring washer from the bearing hub, and hold it against the non-drive-end bearing on the rotor shaft. Does the wave spring washer make contact with only the outer race of the rotor bearing?

Step 7

Remove the drive-end end bell from the motor housing, as shown in Figure 3-12. Looking inside the drive-end of the motor with the end bell removed, notice that if the motor has an internal cooling fan, it will likely be on this end of the motor.

Notice also that the drive-end end bell also has a bearing hub to hold the rotor bearing on this end of the motor. Compare the size of the rotor bearing on the non-drive-end of the rotor shaft with the rotor bearing on the drive-end of the shaft, and note that the drive-end side bearing is probably larger. The reason for making the drive-end bearing larger is because it takes more abuse from forces such as belt tension if the motor is driving a pulley-and-belt application. The non-drive-end bearing just keeps that end of the rotor shaft centered in the stator, and does not experience the same forces. For motors with smaller fractional horsepower, the bearings may be the same size on both ends, but motors with larger horsepower will have different-size bearings.

Figure 3-12 Drive-end end bell removed

1. Is the drive-end bearing larger than the non-drive-end bearing on the motor you are using?

Step 8

Carefully remove the rotor from the stator, if it did not already come out with the drive-end end bell, and look through the motor stator core. Depending on the design of the motor you are using, the auxiliary start switch contact lever and the auxiliary start switch contact may or may not be visible. If the contact assembly is mounted to the motor stator, as is the case with this motor, the auxiliary start switch contact lever and auxiliary start switch contact will be visible, as shown in Figure 3-13. Some motor designs, however, mount the contact assembly to the non-drive-end end bell. If the motor you are using has the contact assembly mounted to the non-drive-end end bell, carefully reattach the non-drive-end end bell with the contact assembly to the non-drive-end of the motor housing; it is not necessary to use any type of fasteners for this.

Now when you look through the stator core you should see the auxiliary start switch contact lever that the shoulder from the centrifugal weight assembly will press against. Note how the auxiliary start switch contact lever is close to the center of the stator core, where it will make contact with the shoulder from the centrifugal weight assembly. Press on the auxiliary start switch contact lever, and note how short the throw is between the open and closed contact states. Connect a meter set for continuity across the auxiliary start switch contacts, and verify that the contacts are open at rest and closed when you press on the contact lever. When the rotor is stopped, the springs on the centrifugal weight assembly pull the weights in, and the shoulder is pushed out against the contact lever, closing the auxiliary start switch contacts and engaging the auxiliary start windings. When the motor starts to turn close to synchronous speed, the weights on the centrifugal weight assembly will be thrown out, which will pull the shoulder in, away from the contact lever, and allow the auxiliary start winding contacts to open and de-energize the auxiliary start windings.

1. Does the motor you are using have the contact assembly mounted to the non-drive-end end bell, the stator core, or the motor housing?
2. Did the auxiliary start switch contacts test open when the switch lever was at rest, and closed when you pressed on the contact lever?

Also, from this view of the motor stator core, it is easy to identify the main run windings and the auxiliary start windings. The main run windings have many turns of larger-gauge wire, and the auxiliary start windings have fewer turns of smaller-gauge wire.

1. Are you able to identify two distinctly different types of coils in the motor you are using?

Figure 3-13 Stator from drive-end without rotor

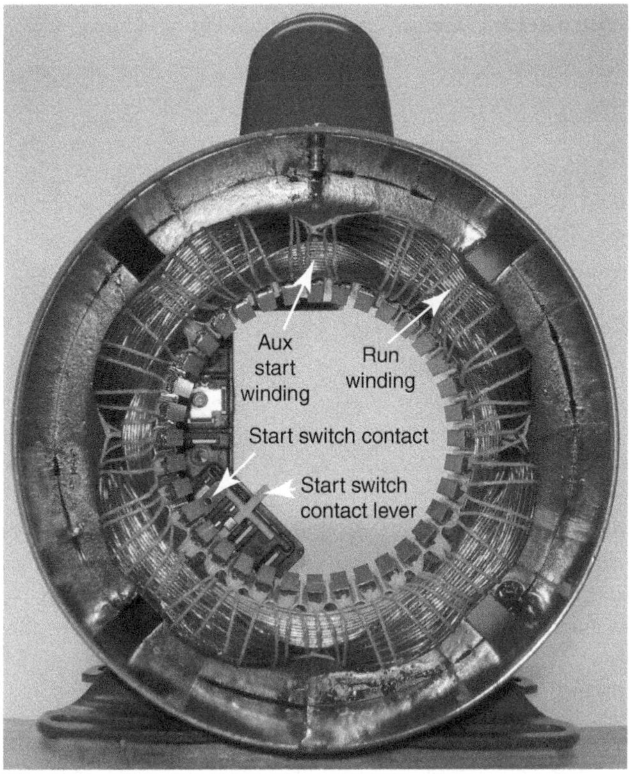

Step 9

Figure 3-14 shows the rotor assembly of the single-phase motor. Look closely at the iron core, and notice that it is constructed with many separate thin iron laminations to reduce eddy currents. Also identify the poured aluminum shorting rings on each end of the iron laminated core, and look around the edge of the shorting rings to see the aluminum rotor bars poured into the cavities through the rotor core.

Figure 3-14 Single-phase motor rotor

Step 10

Figure 3-15 shows the centrifugal weight assembly on the non-drive-end of the rotor shaft. The rotor is not turning, so the spring tension of the spring across the weights is holding them in, which pushes the weight assembly shoulder out. In this position the centrifugal weight assembly shoulder would be pressing against the auxiliary start winding switch contact lever, holding the auxiliary start winding contacts closed, and ready to energize the auxiliary start winding when the motor is energized.

Figure 3-15 Centrifugal weight assembly weights pulled in

Figure 3-16 shows the centrifugal weight assembly with the weights in the thrown-out position. When the motor is energized, and the rotor speed approaches synchronous speed, the spring tension of the spring across the weights will no longer be able to hold them in and they will be thrown out into this position. Notice that when the weights were thrown out, the weight assembly shoulder was pulled in. This movement of the weight assembly shoulder pulls it away from the auxiliary start switch contact lever, and allows the auxiliary start switch contacts to open, de-energizing the auxiliary start winding.

With your fingers, gently pry the centrifugal weight assembly weights apart, and note how the assembly's shoulder moves in and out to make and break contact with the auxiliary start switch contact lever.

1. Does the shoulder on the centrifugal weight assembly move back and forth on the rotor shaft as the weights are moved in and out?

Figure 3-16 Centrifugal weight assembly weights thrown out

Step 11

Remove the fasteners holding the start capacitor doghouse on top of the motor, and gently turn it over to reveal the capacitor inside, as shown in Figure 3-16. Start capacitors are easy to distinguish from run capacitors, because they have very high microfarad values—almost always triple-digit values—whereas run capacitors have single- and double-digit values. Start capacitors are AC electrolytic capacitors with the hard Bakelite package and the paper-covered blowhole in the top, whereas run capacitors are paper and foil with an aluminum package.

Figure 3-17 Start capacitor dog house

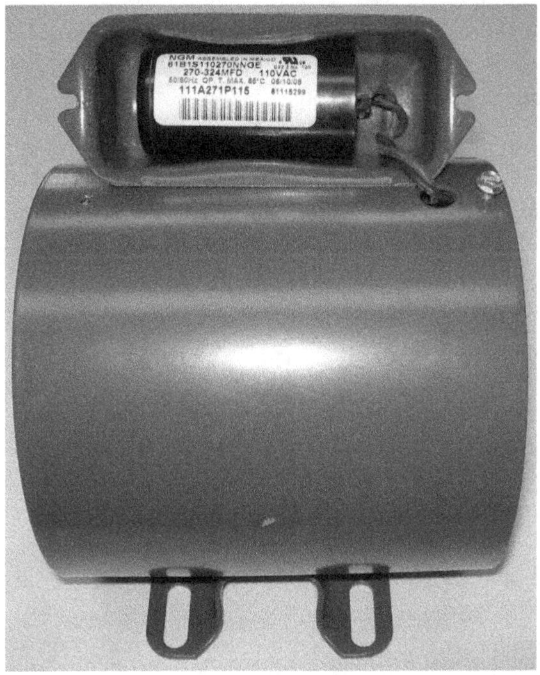

1. What is the microfarad rating of the start capacitor in the motor you are using?
2. What is the voltage rating of the start capacitor in the motor you are using?
3. Is the start capacitor in the motor you are using rated for AC?
   a. What Hz?
4. Does the capacitor in the motor you are using have a blowhole in the top?

THREE-PHASE MOTORS

Step 12

Acquire a three-phase motor to disassemble. Figure 3-18 shows a three-phase motor, which will be used for demonstration purposes here; the motor you use may look much different. Disassembling the three-phase motor is basically the same as disassembling the single-phase motor, so those instructions are not going to be repeated. The focus for the three-phase motor, instead, is on the differences of the rotor and stator from those of the single-phase motor.

Figure 3-18 Three-phase motor

Step 13

Disassemble the three-phase motor and remove the rotor; it should resemble the rotor shown in Figure 3-19. Note immediately that the iron core of a three-phase motor is very similar to the single-phase motor rotor. The iron core is made of very thin iron laminations stacked together, molten aluminum is poured into the rotor bar cavities in the iron core, and aluminum shorting rings are poured on each end.

The three-phase motor rotor does not have a centrifugal weight assembly on the rotor shaft, because three-phase motors do not have auxiliary start windings that need to be de-energized once the motor approaches synchronous speed. This particular motor also does not have a cooling fan, because it is a totally enclosed, external-fan-cooled motor; the cooling fan is on the outside of the motor on the non-drive-end.

Figure 3-19 Three-phase motor rotor

The three-phase motor rotor also has two bearings on the shaft; the non-drive-end bearing is even with the end of the shaft, as shown in Figure 3-20, and the shaft protrudes beyond the bearing on the drive-end, as shown in Figure 3-21.

Figure 3-20 Three-phase motor rotor, non-drive-end

Figure 3-21 Three-phase motor rotor, drive-end

Step 14

Look through the stator core of the three-phase motor, as shown in Figure 3-22. Note the absence of an auxiliary start winding contact lever, or auxiliary start winding contacts, again because the three-phase motor does not have auxiliary start windings. The only wear points on a three-phase motor are the bearings.

Figure 3-22 Three-phase motor stator

Look closely at the stator windings of the three-phase motor, as shown in Figure 3-23, and notice that there is only one type of winding—the main run windings. In the single-phase motor it was easy to identify two different types of winding coils, the main run windings and the auxiliary start windings, but three-phase motors do not have auxiliary start windings.

Figure 3-23 Three-phase stator windings

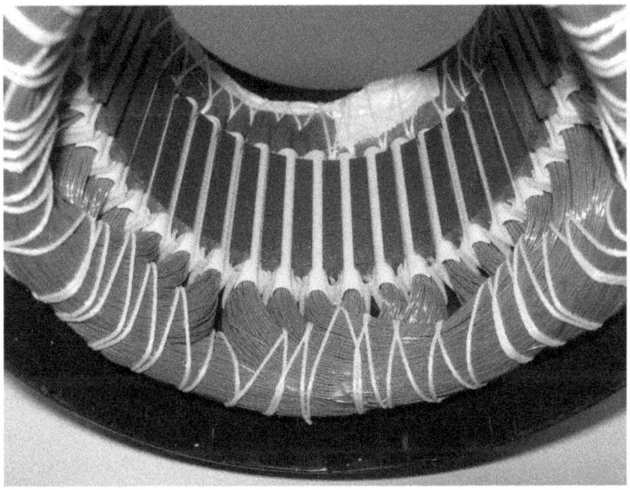

Step 15

After making all of your observations about these motor component parts, reassemble the motors and put them away.

QUESTIONS

1. What is the reason given in the text for replacing smaller electric motors rather than repairing them?
2. Why do some customers send their motors to a motor repair shop rather than having electricians do the repair work, even for simple repairs such as bearings?
3. What is one advantage of making a simple motor repair, such as replacing bearings, in the field?
4. Why is it important to mark both end bells of the motor with the motor housing before disassembling the motor?
5. Why do motors have grounding provisions inside the terminal box?
6. On which end of the rotor shaft is the centrifugal weight assembly, the drive end or the non-drive end?
7. Why must the outer race of the bearing fit snugly into the bearing hub in each of the motor end bells?
8. The non-drive-end end bell has a special washer in the bearing hub. What is it called, and what is its purpose?
9. If a motor has an internal cooling fan to circulate air inside the motor, which end of the motor will it likely be on?
10. What is the reason that the drive-end bearing would be larger than the non-drive-end bearing?
11. When the motor is at rest, are the centrifugal weight assembly weights in the in or out position?
12. With the centrifugal assembly weights in this position, is the assembly's shoulder pressing or not pressing the auxiliary switch contact lever?
13. With the centrifugal assembly weights in this position, are the auxiliary start winding contacts open or closed?
14. In the single-phase motor stator, which winding coils have the fewest turns of the smaller wire?
15. In the single-phase motor stator, which winding coils have the most turns of the larger wire?
16. What is the purpose of the springs holding the weights of the centrifugal weight assembly pulled in?
17. How does the microfarad value of a start capacitor compare with that of a run capacitor?
18. Because AC electrolytic start capacitors depend on a chemical reaction that produces heat when they are in an energized circuit, what provision do they have to prevent them from exploding?
19. Why don't three-phase motors have centrifugal weight assemblies and auxiliary start switches?
20. What are the only two wear points of three-phase motors?
21. What is the main difference between single-phase and three-phase motor stator winding coils?

LAB EXERCISE #4

Single- and Three-Phase Power

Name:	Date:	Grade:

Purpose

The purpose of this lab is to familiarize the learner with the concepts and terminology associated with single- and three-phase electric power circuits, and to practice taking meter readings of the electrical parameters in each.

Materials:

Digital clamp meter

3–120-V lamps in sockets

SINGLE- AND THREE-PHASE POWER

With the exception of some specialized direct current (DC) applications, such as electrolysis plating and charging systems, most electrical energy systems that construction electricians work with are alternating current (AC). When electrical energy is produced electromechanically from a rotating machine, the resulting waveform is a sine wave, as shown in Figure 4-1. The frequency of the alternating current is measured in cycles per second, and is labeled as Hz. Notice in Figure 4-1 that the horizontal axis is labeled "Time," which may also be expressed in degrees, and the vertical axis is labeled "Magnitude." The frequency of the electrical energy supply in North America is 60 Hz, meaning that 60 cycles of the energy source voltage polarity are completed every second; the frequency is 50 Hz in most European countries.

Figure 4-1 Single-phase sine wave

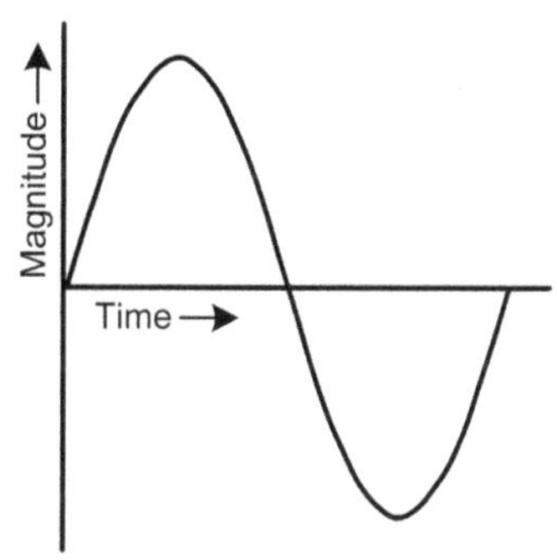

One cycle is defined as the time from one point on a waveform to a second point on the same waveform where it starts to repeat. Figure 4-2 shows a sine wave with several different numbered points along the sine wave waveform to help understand what a cycle is. From each point with a number to the second point with the same number is a complete cycle, because the waveform starts to repeat. Just to make certain you understand the cycle concept, examine the point labeled 4 on the waveform, which does not represent a cycle. Mark your own number 4 on the waveform where it should be to mark a full cycle.

Figure 4-2 Defining a cycle

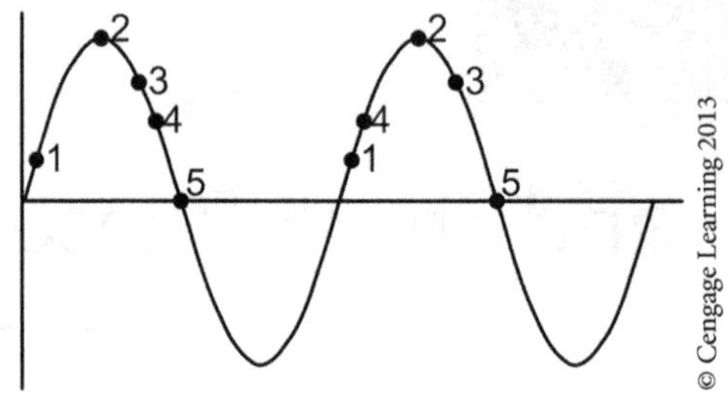

It is important to understand how to determine a complete cycle as facility electrical systems become more and more troubled by harmonic currents. Harmonic currents in an electrical system can deform the waveform, and waveforms similar to the one simulated in Figure 4-3 are possible in severe cases. If a cycle is less stringently defined, such as, "from the time that the waveform goes positive, and then negative, and then starts going positive again," it could cause problems in some situations. Harmonics is an ever-increasing problem in electrical systems, and any commercial or industrial electrician in the trade today will encounter problems caused by harmonics.

Figure 4-3 Harmonic distorted waveform

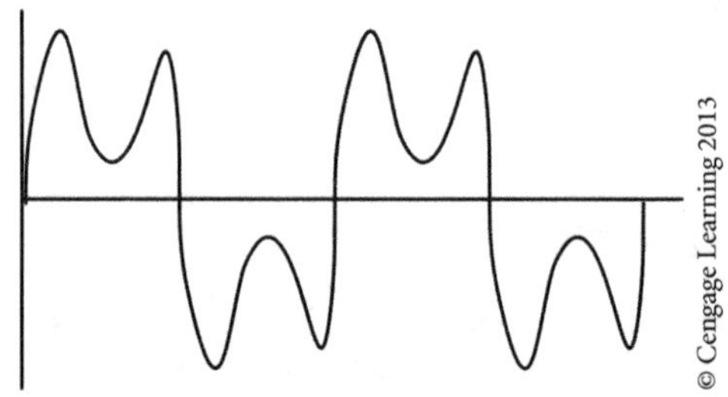

The vertical axis in Figure 4-1 is labeled "Magnitude," which may represent many different electrical parameters, but the focus here is on voltage and current. As electricians, we are concerned with the root-mean-square (RMS) value of magnitude measurements, because the test meters we use display that value. Figure 4-4 shows the full sine wave of one complete cycle with both the RMS and peak values labeled in both the positive- and negative-magnitude polarities. As electricians, whenever an AC voltage or current value is not labeled with an identifier such as *average*, *RMS*, or *peak*, it is assumed to be an RMS value.

The significance of the RMS value is that it represents the DC equivalent value of a sine wave. What does DC equivalence mean? DC voltage does not fluctuate with time; it is steady. AC voltage, however, fluctuates constantly. For the points along the horizontal time axis of the sine wave that the voltage is below the RMS line, there have to be other times that it is above the RMS line in order to average out to the constant DC level over time. Simply stated, in an AC circuit the RMS value of a constantly changing sine wave waveform will produce the same amount of power over the time period of a cycle as the steady DC value of the same value.

Figure 4-4 Single-phase sine wave with labels

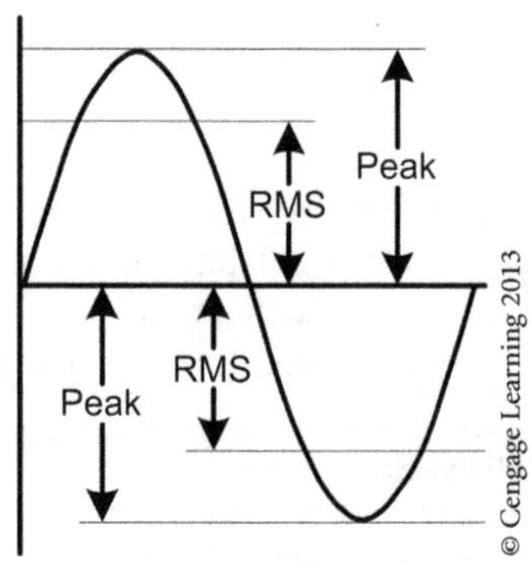

If the AC waveform is a pure sine wave, the RMS value may be calculated by multiplying the peak magnitude value times 0.707, as shown in Figure 4-5. The multiplier is not valid if any harmonics are present, in which case a true RMS reading meter would be needed for an accurate reading. The true RMS reading meter will actually measure the deformed waveform, and calculate the RMS value. Non-true RMS reading meters will only measure the peak value and apply the 0.707 multiplier.

Figure 4-5 Single-phase sine wave with values

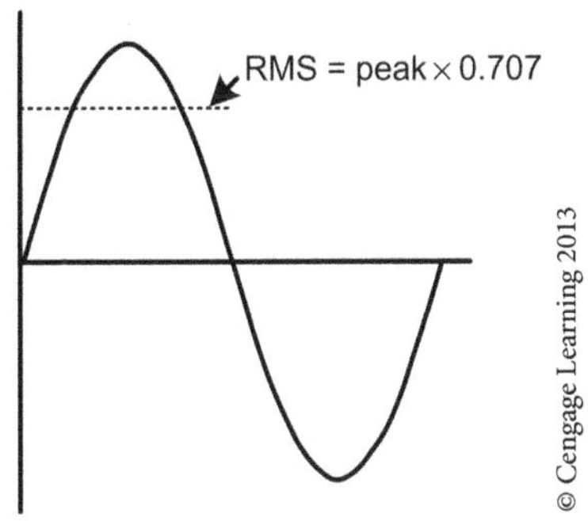

Figure 4-6 shows that the normal 120-V_{RMS} branch circuit actually has a peak voltage of approximately 170 volts. Figure 4-7 shows an actual oscilloscope line-grid waveform display for a 120-VAC circuit measurement, with a horizontal dotted line added to mark where 120 volts is on the vertical magnitude scale. Notice that the peak of the waveform is considerably higher than the 120-volt level; it is approximately 170 volts peak.

Figure 4-6 Single-phase sine wave with instantaneous values

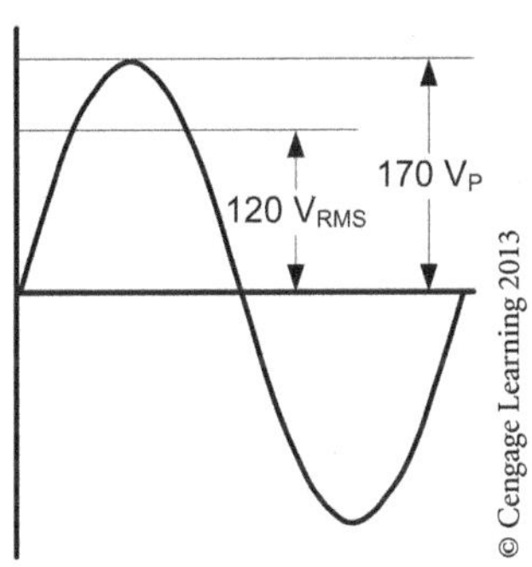

Figure 4-7 Oscilloscope voltage waveform

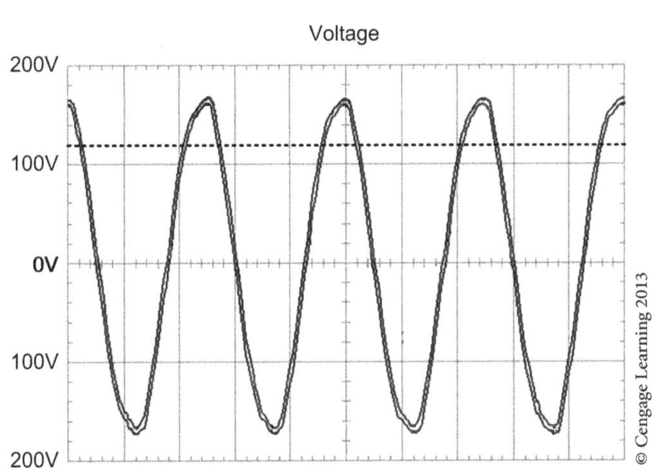

The readings demonstrated in Figures 4-6 and 4-7 correspond with the voltage that would be read across the X1-X2 coil in Figure 4-8, from one ungrounded conductor to the grounded center-tap point between the transformer coils. Notice in Figure 4-8 that the line-to-line voltage reading is 240 volts, twice the line-to-center-tap grounded point. The line-to-line voltage is doubled because the two coils are 180° out-of-phase with each other, and the voltages of each individual coil will add directly. This single-phase configuration is the same type of power source found in residential electrical services.

Figure 4-8 Single-phase transformer connection

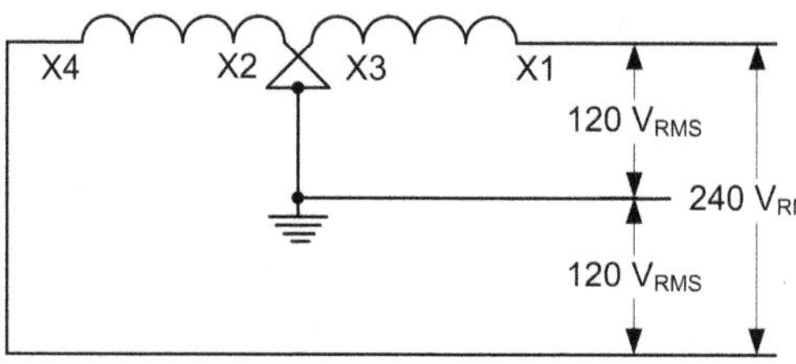

SINGLE-PHASE LAB METER READINGS

The purpose of this section of the lab is to demonstrate that there are two different kinds of single-phase loads: single-phase line to neutral and single-phase line to line. Line-to-neutral loads are considered unbalanced loads, because only half of the transformer secondary of the power supply is used. Line-to-line loads are considered balanced, because both halves of the transformer secondary of the power supply are used. Unbalanced loads require a neutral connection, whereas balanced loads do not.

Step 1

Locate a single-phase 120/240-volt power source, and complete the voltage readings for Table 4-1.

Single-Phase 120/240 Volt		
L1 to N	L2 to N	L1 to L2

Table 4-1 Single-phase 120/240-volt readings

Step 2

Connect the unequal lamp circuit shown in Figure 4-9. Before taking any current readings, energize the circuit and unscrew each lamp, one at a time, to prove that either lamp will work independently of the other.

1. Do the lamps work indepently of one another with the neutral connection?

Figure 4-9 Single-phase unequal lamp circuit

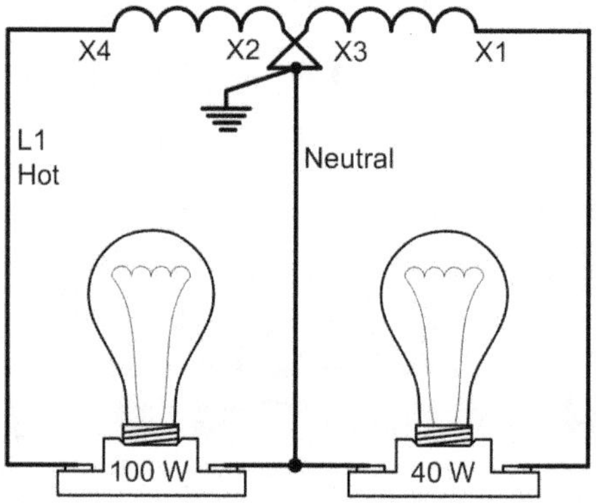

Step 3

Using your clamp meter, make the current readings as shown in Figure 4-10 to complete Table 4-2.

Figure 4-10 Single-phase circuit meter readings

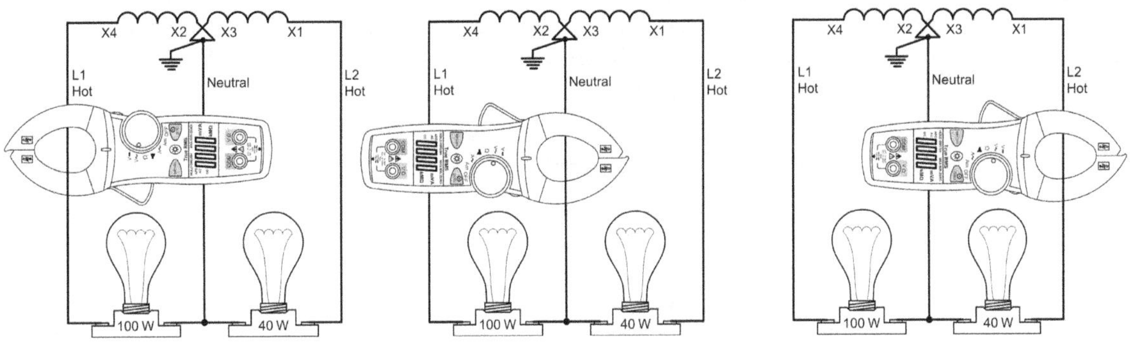

Unbalanced Lamp Current Readings		
L1 100 W Hot	Neutral	L2 40 W Hot

Table 4-2 Single-phase unbalanced lamp current readings

1. Did the neutral conductor in this circuit have a current reading?
2. Was the current reading equal to the difference between the L1 and L2 readings?
3. With the circuit energized and the clamp meter on the neutral conductor, unscrew the 40-watt lamp. Does the neutral current decrease, remain the same, or increase?
4. Explain how removing a load from the circuit in the last step caused the neutral current to increase.
5. Do not do it, but could the neutral conductor in this circuit be removed without affecting the circuit operation?

Step 4

Replace the 40-watt lamp in this circuit with a second 100-watt lamp, so that the two lamps are of equal size. Repeat the current readings as shown in Figure 4-10 to complete Table 4-3.

Balanced Lamp Current Readings		
L1 100 W Hot	Neutral	L2 100 W Hot

Table 4-3 Single-phase equal lamp current readings

1. Did the neutral conductor in this circuit have a current reading?
2. Are the current readings between each of the L1 and L2 conductors equal?
3. Could the neutral conductor in this circuit be removed without affecting the circuit operation?
4. Remove the neutral conductor and energize the circuit. Unscrew one of the 100-watt lamps. Why did the second lamp also go out?

This circuit with the two 100-watt lamps is actually operating like a single 240-volt line-to-line load. The lamps are of equal resistance, so they will have equal voltage drops—one-half of the 240-volt supply voltage.

THREE-PHASE POWER

The purpose of this section of the lab is to apply the concepts and terminology specific to three-phase power to prove them in the lab environment.

Three-phase power is really nothing more than having three sets of stator coils in the alternator for each cycle, rather than only one set of stator coils for single-phase power. For each cycle, three sine wave waveforms are created instead of only one, as shown in Figure 4-11. The three sets of stator coils can be connected into two different three-phase configurations: wye or delta.

Figure 4-11 Three-phase sine waves

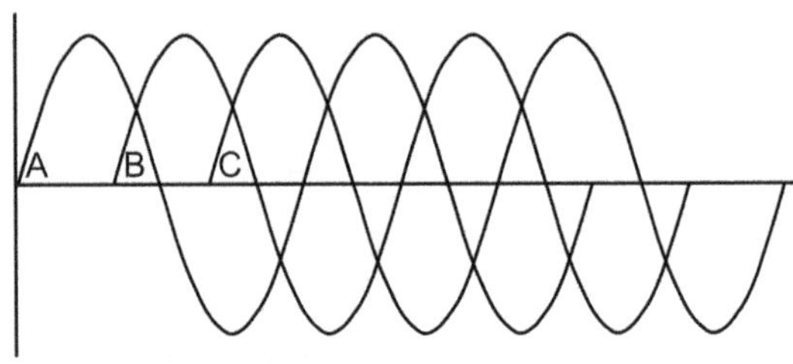

Figure 4-12 shows how the wye and delta connections are configured. The wye configuration is named for its resemblance to the English letter Y, and the delta is named for its resemblance to the greek letter delta. The first terminology that must be understood is the phase and line relationship between the wye and delta configurations. As summarized in Table 4-4, in the wye configuration, the phase and line dimensions are different; the line and phase currents are equal, but the line and phase voltages are different. In the delta configuration, the phase and line dimensions are the same; the line and coil voltages are equal, but the line and phase currents are different.

Figure 4-12 Three-phase definitions

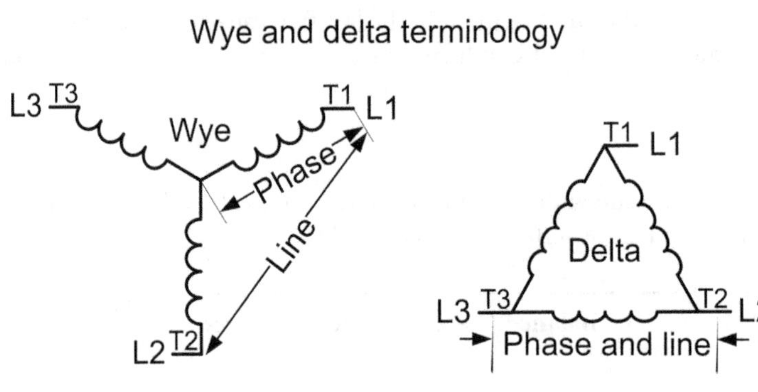

Wye and Delta Line and Phase Characteristics	
Wye	Delta
$E_L = E_P \times 1.732$	$E_L = E_P$
$I_L = I_P$	$I_L = I_P \times 1.732$

Table 4-4 Line and phase characteristics

Figure 4-13 shows a wye-connected transformer, with the phase and line voltages labeled with both peak and RMS values. Reading from one line to the center grounded point where one end of all the phases connect for the wye configuration, the voltage reading is 120 V_{RMS}. When a voltage measurement is taken from line to line on a wye-configured transformer, the voltage is not double 120 V_{RMS}; it is 208 V_{RMS}. The reason the line-to-line voltage was doubled in the single-phase configuration in Figure 4-8 is because the transformer coils are 180° out-of-phase. In Figure 4-13, the three-phase wye configuration, the transformer coils are 120° out-of-phase. To calculate the 120/208-volt values for this configuration, the 1.732 multiplier, which is the square root of 3, is used. Using the formula $E_{Line} = E_{Phase} \times 1.732$, then $E_{Line} = 120\ V \times 1.732$, or 208 V.

Figure 4-13 Three-phase transformer connection

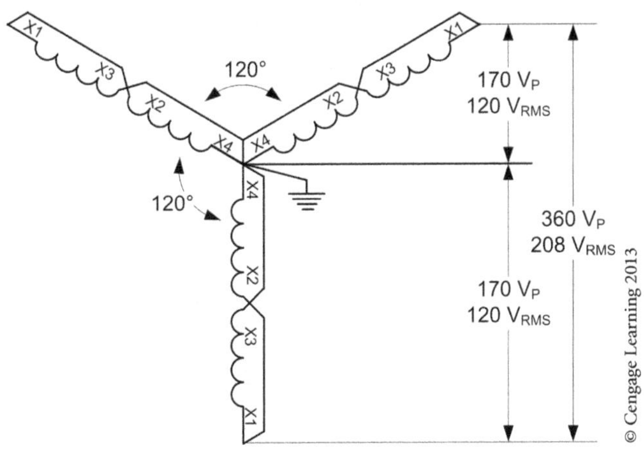

To make the 120/208-volt values easier to visualize, Figure 4-14 shows the horizontal axis with only two of the phase voltage waveforms the reading is taken between, rather than all three of the phases, to reduce the clutter. What may make this graphic harder to follow is that the reading is taken between two phases, rather than between a phase and the grounded center point of the configuration. This time, rather than reading from one of the sine wave waveforms to the horizontal zero reference line, the reading will be taken between the two waveforms. To demonstrate how the 208 volts is derived, all calculations in this diagram must be made with peak and instantaneous voltage readings:

$$E_{Peak} \text{ and } E_{Instantaneous} \cdot E_I = \text{Sin } \angle\theta \times E_P.$$

First of all, the two sine wave waveforms for phases A and B are drawn 120° out-of-phase on the horizontal axis of the drawing. Look at the resultant waveform on each side where it is touching the zero-magnitude horizontal reference line. Follow the dotted lines up, and note that the phase A and B waveforms are together at these times; there is no voltage difference between them. Where the resultant waveform peaks is 120° from both phases A and B, as noted by the 120° dimensioning arrows in the drawing. The formula for finding the instantaneous voltage at the peak of the resultant waveform is:

$$E_I = \text{Sin } \angle\theta \times E_P \quad = \quad E_I = \text{Sin } 120° \times 170\ V \quad = \quad 147\ V_I$$

The calculation for the instantaneous voltage of the positive-polarity phase B waveform 120° from its start point to the horizontal reference line is 147 V_I. The calculation for the instantaneous voltage of the negative-polarity phase A waveform 120° from its end point to the horizontal reference line is also 147 V_I. As shown in the drawing, the two voltages added 180° out-of-phase equal 294 V_I. Instantaneous voltage is calculated from peak voltage, so the 294 V_I must be multiplied by 0.707 to calculate the RMS value of 208 volts, line to line.

Figure 4-14 Line-to-line voltage explanation

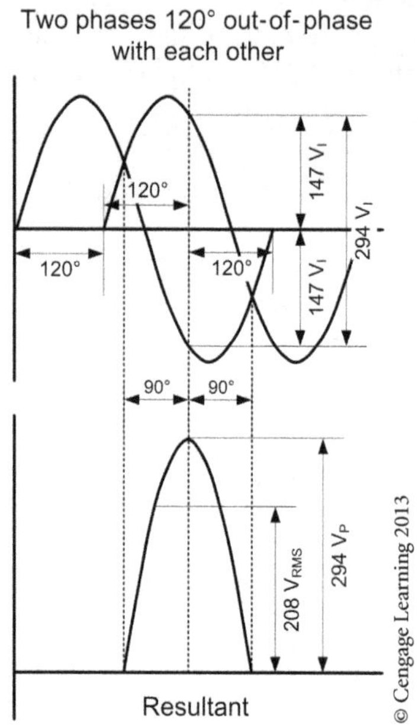

LAB FOR THREE-PHASE METER READINGS

Step 5

Locate a three-phase 120/208-volt wye power source, and complete the voltage readings for Table 4-5.

Three-Phase Wye Voltage Readings		
L1 to Neutral	L2 to Neutral	L3 to Neutral
L1 to L2	L2 to L3	L3 to L1

Table 4-5 Three-phase wye voltage readings

1. Are the phase-voltage readings and the line-voltage readings different?
2. Are the readings different by the three-phase multiplier, 1.732?

Step 6

Figure 4-15 shows the wye configuration connection with three lamps, first in the traditional wye shape, and then connected with the lamps lined up in a straight line. In the drawing with the lamps in the wye shape, notice that one side of each of three lamps is connected together at a common point, and then lines 1, 2, and 3 connect to the other ends of the lamps. The drawing with the lamps lined up straight is the same drawing. Connect this circuit and complete the current readings for Table 4-6.

Figure 4-15 Wye lamp connection

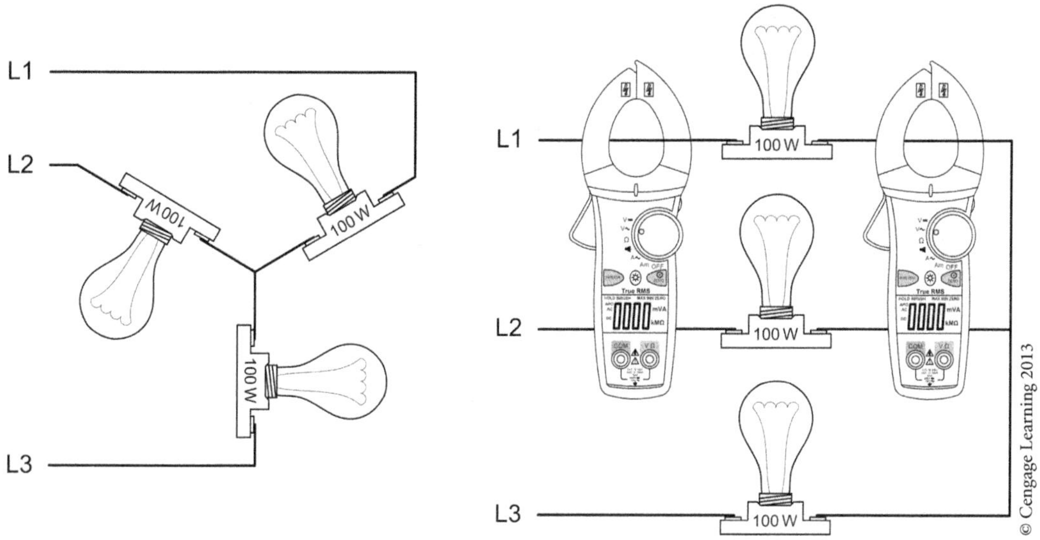

If all you are able to locate is a 480-volt wye power source, the drawing in Figure 4-15 will have to be modified to install two lamps per phase, as shown in Figure 4-16, because the phase voltage is going to be 277 volts. This is still overvoltage for the lamps, but it should work long enough to take the current readings without damaging the lamps. If you are concerned, install three lamps per phase; they will just not burn as brightly.

Figure 4-16 Wye 480-V connection

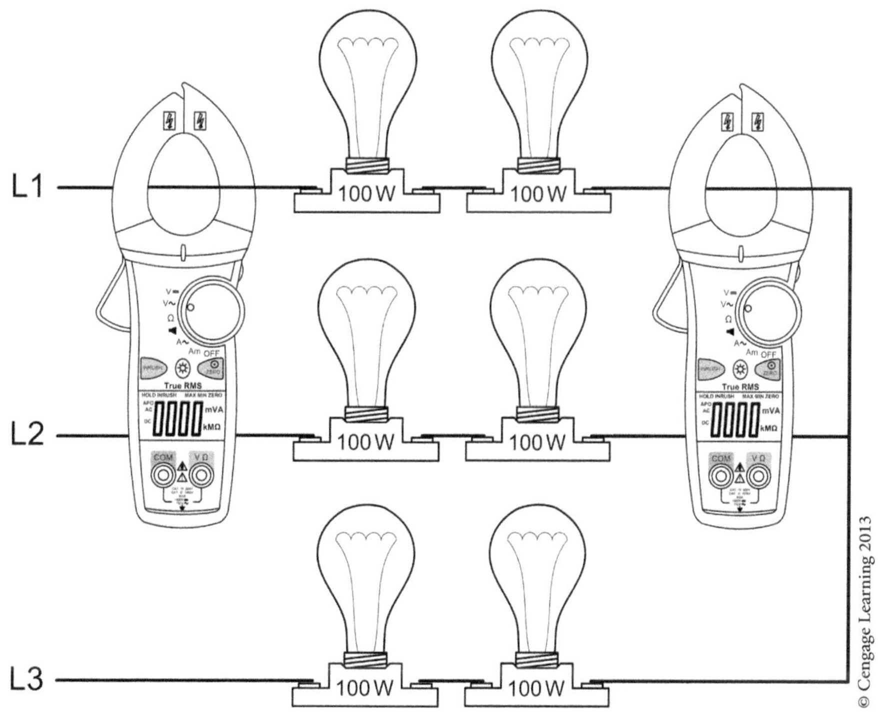

Three-Phase Wye Current Readings		
L1	L2	L3
Phase 1	Phase 2	Phase 3

Table 4-6 Three-phase wye current readings

1. Are all of the current readings the same?

Step 7

Locate a three-phase 240-volt delta power source, and complete the voltage readings for Table 4-7.

Three-Phase Delta Voltage Readings		
L1 to L2	L2 to L3	L3 to L1

Table 4-7 Three-phase delta voltage readings

1. Are all of the voltage readings the same?

Step 8

Connecting the delta configuration using 120-volt lamps for the load requires that two lamps are connected in series for each phase, as shown in Figure 4-17, because the phase voltage equals the line voltage. Connect the circuit and complete the current readings for Table 4-8. Meter A shows an example of a line-current reading, L1. Meter B shows an example of a phase-current reading, phase 2.

Figure 4-17 Delta lamp connection

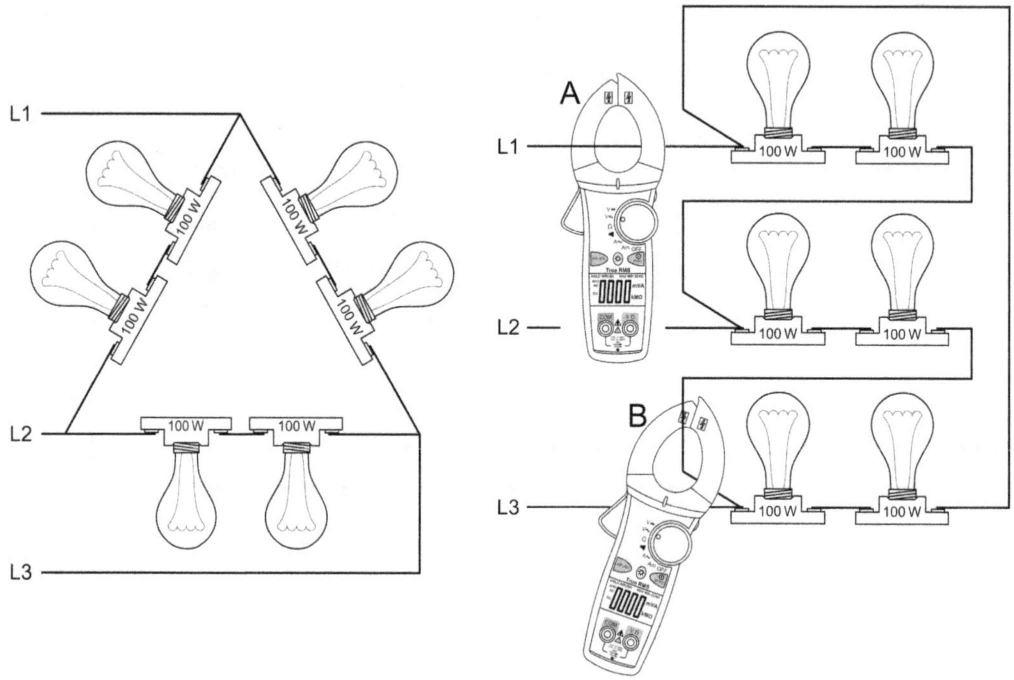

If all you are able to locate is a 480-volt delta power source, the drawing in Figure 4-17 will have to be modified to install four lamps per phase to accommodate the 480 volts.

Three-Phase Delta Current Readings		
L1	L2	L3
Phase 1	Phase 2	Phase 3

Table 4-8 Three-phase delta current readings

1. Are the phase-current readings and the line-current readings different?
2. Are the line- and phase-current readings different by the three-phase multiplier, 1.732?

An easy method to remember if the 1.732 three-phase multiplier is applied to the voltage or current in a wye or delta configuration, is to recall the basic rules of series and parallel circuits. In a series circuit current is the same in all parts, but the total voltage is the sum of all the circuit component voltage drops. In a parallel circuit the voltage is the same in all parts, but the circuit current is the sum of all the current paths.

Figure 4-18 shows how to think of the wye configuration as a series circuit, because each line current has a single path through the load connected to that line; and the delta configuration as a parallel circuit, because each line current is split between two different load current paths. Wye, series, current is the same, so apply the multiplier to the current. Delta, parallel, voltage is the same, so apply the multiplier to the current.

Figure 4-18 Wye and delta current paths

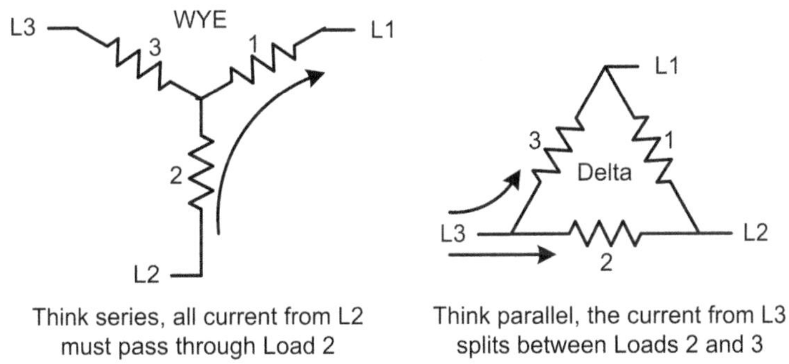

QUESTIONS

1. When drawing the two-dimensional X and Y axes to record a sine wave, as in Figure 4-1, how is each axis labeled?

   Horizontal: ______________________ Vertical: ______________________

2. What is the definition of a complete cycle on an AC waveform?
3. What may happen to the sine wave waveform if harmonic currents are present?
4. If an AC voltage or current value is not labeled with an identifier such as average, RMS, or peak, what value should it be assumed to be?
5. An oscilloscope reads peak voltage. If the waveform is a pure sine wave, what multiplier is used to calculate the RMS value?
6. Why are the voltage values of multi-voltage, single-phase systems even multiples of the lower value (E × 120/240 V)?

7. Is a line-to-neutral load in a single-phase system a balanced or unbalanced load?
8. Which single-phase loads require a neutral connection, balanced or unbalanced?
9. What are the two different configurations that three-phase circuits may be connected in?
10. How many degrees are the phases of a three-phase system separated by?
11. What is the three-phase multiplier, which is also the square root of 3?
12. To help remember if the 1.732 three-phase multiplier is applied to the voltage or current, how may the wye and delta configurations be thought of?
13. If three individual 1-kW, 120-V heater elements are to be connected in a three-phase wye configuration, what is the minimum current rating that would be required for the line conductors on a 120/208-V system?
14. If three individual 1-kW heaters were rated for 240 V, and were connected to a 240-V three-phase delta system, what is the minimum current rating that would be required for the line conductors?
15. If three individual 1-kW heaters were connected in a wye configuration to a 480-V line-to-line wye system, what voltage would each heater need to be rated for?
16. If three individual 1-kW heaters were connected in a delta configuration to a 480-V line-to-line delta system, what voltage would each heater need to be rated for?

LAB EXERCISE #5

Single- and Three-Phase Motor Continuity Tests

Name:	Date:	Grade:

Purpose

The purpose of this lab is to practice taking continuity readings of single- and three-phase induction motors, including basic ohmmeter tests that should be performed with all new motors when they arrive on the jobsite to be certain they were not damaged in shipping.

Materials:

Single-phase motor with the auxiliary start and main run windings separately accessible

Three-phase motor that is internally connected in a wye configuration

Three-phase motor that is internally connected in a delta configuration

Clamp meter, tested and verified for proper operation

TESTING FOR SHIPPING DAMAGE

Electric motor manufacturers perform extensive quality assurance testing on their products to guarantee that the consumer receives a product free of manufacturing defects. These quality assurance tests are much more sophisticated and tailored to the individual motor than those tests the electrician can effectively perform in the field, but that does not preclude all testing by the electrician. Even though it is rare to find a defective motor in the field because of manufacturing defects, it is not uncommon to find a defective motor in the field because of shipping damage.

Although it is less important to test smaller, inexpensive motors when they arrive on the jobsite, it is very important to test larger, more expensive motors to affix responsibility for any damage. When a motor is received on a jobsite, it may be weeks or months before it is actually installed and tested. In such cases, if the motor was damaged in shipping, it will be very difficult to prove the damage happened during shipping, and not when the motor was moved around the jobsite. In these cases the electrical contractor will likely incur the cost of replacing the motor, and the problem could have been avoided if the motor had been tested when it first arrived on the jobsite.

It does not require specialized test equipment to test a motor when it arrives on a jobsite, because the type of damage caused by a shipping accident is usually severe enough that a simple continuity test will identify the problem. Before the person delivering the motor leaves the jobsite, the technician should take a few minutes to perform a visual inspection for obvious external physical damage, check for the presence of a drop shock indicator to see if the motor has been mishandled, and open the motor terminal box and perform a simple continuity test. If any problems are discovered, it is best to have the delivery person document them on the paperwork and sign it, and let the purchasing department settle the dispute with the shipping company. Once the technician's signature is on the paperwork as having received the motor, the contractor owns it.

MEGGER TESTING

Some electricians argue that the only way to test a motor is with a megger, but one is not always available on every jobsite. A megger is a specialized meter that tests the conductor insulation of motors, or other electrical equipmen by applying a higher DC voltage to the circuit than a normal multimeter set for resistance or continuity would. Common megger voltages are 300 V, 750 V, and 1,000 V, which certainly does stress the insulation more than the 9 volts of a multimeter—but is this necessary? As mentioned, new motors have already been tested by the manufacturer, and any damage caused by shipping will normally be severe enough to be found with a simple continuity check. If a megger is available, great! If not, even for a megger advocate, a continuity test is much better than no test at all.

SINGLE-PHASE MOTOR CONTINUITY TEST

Step 1: Verify your test leads and meter

With your meter set in the resistance mode, touch the two meter leads together. With the meter leads touched together, there is no resistance in the meter circuit to read. The display should read "0 Ω," indicating that the meter leads are good and the meter is functioning properly. Zero ohms is the absence of resistance in the meter circuit, a short circuit; "OL" ohms is infinite resistance, an open circuit. OL means that the measured resistance is too large for the meter to register, and the meter is overloaded. Separate the two meter leads so that there is an open circuit, no continuity between the test probes, and verify that the meter display reads "OL."

This lab is going to assume that the motor leads for the auxiliary start and main run windings of the motor you are using for the lab are brought out to a terminal block, similar to what is shown in Figure 5-1. If the motor you are using for this lab does not have the individual motor leads brought out to a terminal block, Table 5-1 identifies the lead conductor colors for you to locate in the motor terminal box.

Note: When making the continuity and resistance measurements for this lab in the motor's terminal box, at least one of the motor leads must be removed from its termination to prevent reading unintentional parallel paths.

Figure 5-1 Single-phase motor terminal block

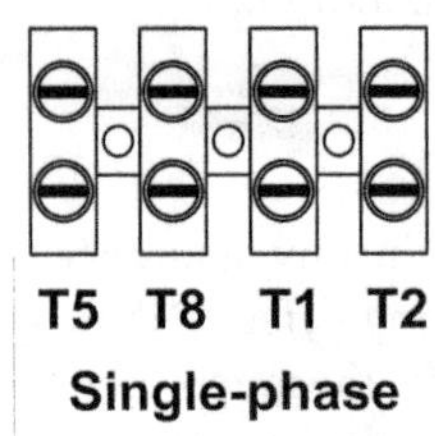

NEMA Single-Phase Lead Colors
T1–Blue
T2–White
T5–Black
T8–Red

Table 5-1 NEMA single-phase motor lead colors

Step 2

The drawing in Figure 5-2 shows a representation of a single-phase motor with the main run windings and the auxiliary start windings brought out to a terminal block. Looking at this diagram, it should be apparent that a continuity check would show continuity between leads T1 and T2, and then between leads T5 and T8. The motor used in this lab is a single-voltage, single-phase motor; if the motor you are using is a dual-voltage, single-phase motor, there will be a second main run winding (leads T3 and T4) that also needs to be tested. There should be nc continuity between any of the T leads and the motor housing, or between the T leads of the main run winding and the T leads of the auxiliary start winding. With your test meter set to continuity, complete the measurements for Table 5-2.

Note: If the motor you are using to make the continuity and resistance measurements for this lab has a start capacitor or a run capacitor, a jumper must be connected across the terminals of the capacitors to take the measurements, or the meter must be connected to the motor stator winding side of the capacitor rather than the terminal in the connection box.

Figure 5-2 Stator coils of a single-phase motor

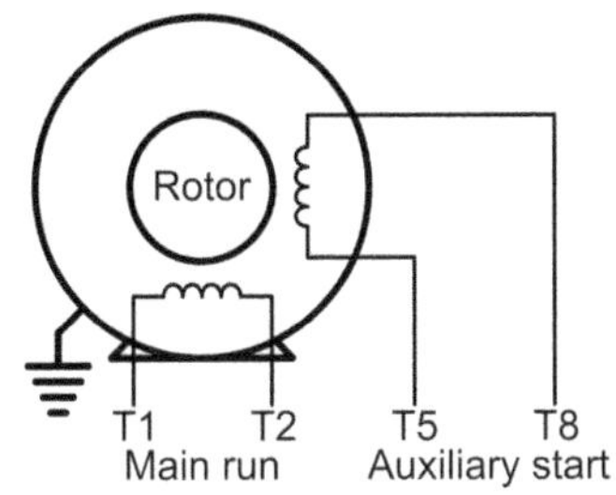

Single-Phase Motor Continuity Test				
T1 to T2	T5 to T8	T1 to T5	T1 to Motor Housing	T5 to Motor Housing
yes Ω	yes	no	no	no

Table 5-2 Single-phase continuity test

1. Were the first two continuity readings, T1 to T2 and T5 to T8, the only continuity readings in the table? Yes
If any of the other reading showed continuity, the motor is damaged; do not use this motor for further testing.

This continuity test of all motor stator coils to the motor housing, without jumping the capacitor, should be performed on motors when they arrive on a jobsite.

Step 3: Megger testing

If you have access to a megger, repeat the same readings between each stator coil, and then each stator coil to the motor housing. Record your readings in Table 5-3.

Single-Phase Motor Continuity Test				
T1 to T2	T5 to T8	T1 to T5	T1 to Motor Housing	T5 to Motor Housing

Table 5-3 Single-phase continuity test

1. Did the measurements in Table 5-3 from the megger display actual measured resistance reading in megaohms (MΩ) from the motor stator coils to the motor housing?

Generally, the megaohm measurements from megger instruments cause more questions and debate than answers when used as a test instrument for motors. Because the megger is capable of reading very high resistance values, it will almost always display a readable resistance of the motor windings to the motor housing of an older motor, so the question becomes, how high does the reading have to be to consider the motor good? You will not have to look very far to find a lot of contradicting rules of thumb based on ohms-per-volt calculations, voltage-class-based minimum resistances that have to be temperature compensated, and even direct answers that range from 0.5 MΩ to 1,000 MΩ.

Megger readings are actually better suited to perform a trend testing, or predictive maintenance, function. When the motor is new, a megger reading is documented and used to set the baseline resistance of the stator coils for a particular motor. Then in subsequent megger tests, the deterioration of the winding insulation may be documented to predict when the motor will ultimately fail, so the motor replacement can be scheduled without causing a disruption in production.

Step 4

In this step you will use the resistance mode of your multimeter to differentiate the auxiliary start winding from the main run winding. With your multimeter set in the resistance mode, make the readings for Table 5-4.

Auxilary Start Winding and Main Run Winding Resistances	
Resistance Reading from T5 to T8	Resistance Reading from T1 to T2
4.3 Ω	1.8 Ω

Table 5-4 Start and run winding resistances

1. Were the two resistance readings of the auxiliary start winding and the main run winding in Table 5-4 different? Yes
2. Which winding had more resistance, the auxiliary start winding or the main run winding? T5-T8
3. Knowing from electrical theory that the smaller the gauge of a wire, the higher the resistance it will have, coupled with knowing that the auxiliary start winding of a motor is wound with a smaller-gauge wire than the main run winding, do the resistance measurements in Table 5-4 make sense? Yes
4. The auxiliary start windings also have fewer turns of wire, which means the coil will have less inductance and therefore less inductive reactance to limit the current when the coil is energized. The auxiliary start winding can only be energized for a short period of time without being damaged from excessive heat. What motor component de-energizes the auxiliary start winding to prevent it from being damaged? T: Switch

THREE-PHASE MOTOR CONTINUITY TESTS

Step 5

The only difference between performing a continuity test on single-phase motors and three-phase motors is understanding the different stator coil configurations—not just between single- and three-phase motors, but between the different three-phase configurations. Three-phase motors are found in different configurations, from three-lead motors to six-, nine-, and 12-lead motors. This lab will focus on the nine-lead wye- and delta-configured three-phase motors.

The nine-lead motor is the most common three-phase motor, because of the versatility of its dual-voltage rating. Each phase of the nine-lead motor has two coil windings. When the two phase coils are connected in parallel, the full-phase voltage will be dropped across each coil, which is for the low-voltage operation of the motor. When the two phase coils are connected in series, half of the phase voltage will be dropped across each coil, which is for the high-voltage operation of the motor. The dual-voltage rating of the motor allows it to be installed in electrical systems with different voltage levels.

Nine-lead, three-phase motors are identical to 12-lead, three-phase motors, but with some internal coil lead connections made at the time of manufacture that are not changeable in the field. Twelve-lead motors may be connected for either wye or delta and high- or low-voltage configuration. Figure 5-3 shows 12-lead motor coils in both wye and delta configurations, but it is to demonstrate how to label the coil leads and to show the internal connections that are made for a nine-lead motor.

Figure 5-3 Twelve-lead wye and delta

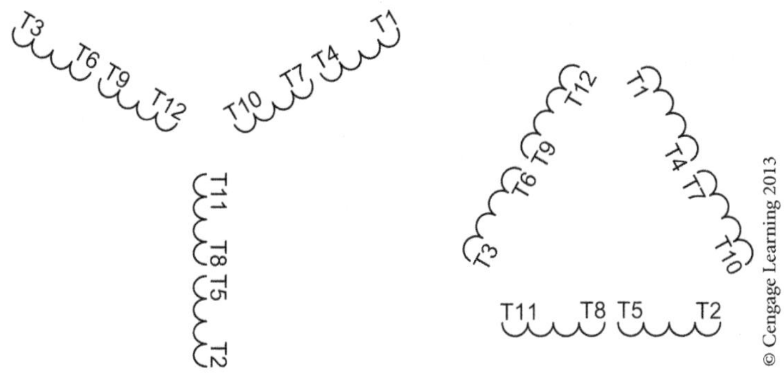

The method for labeling the 12-lead motor coils for the wye-configuration connection is shown in Figure 5-4. Starting on the outer end of any phase coil, label it T1. Turning in a clockwise direction, sequentially label the outer ends of the other two phase coils, T2 and T3. Continuing in a clockwise direction, move to the other end of the first phase coil you labeled T1, and label it T4. Continuing to turn in a clockwise direction, sequentially label the rest of the coil ends while moving closer to the center of the wye.

Figure 5-4 Labeling the leads of a wye-connected motor

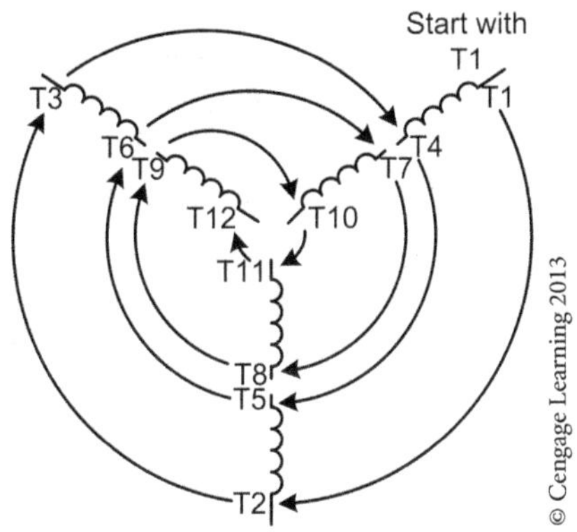

Figure 5-5 shows that the nine-lead wye motor is labeled in exactly the same way, only the coil leads T10, T11, and T12 are internally connected in the motor and are not brought out to the motor terminal box. Even with the internal connection of the three coil leads the motor still retains its dual-voltage rating. The manufacturer connects the T10, T11, and T12 motor leads internally to reduce the number of leads in the motor terminal box. This helps reduce wiring mistakes in the field as fewer wire connections are made, and it reduces installation time.

Figure 5-5 Labeling the leads of a nine-lead wye-connected motor

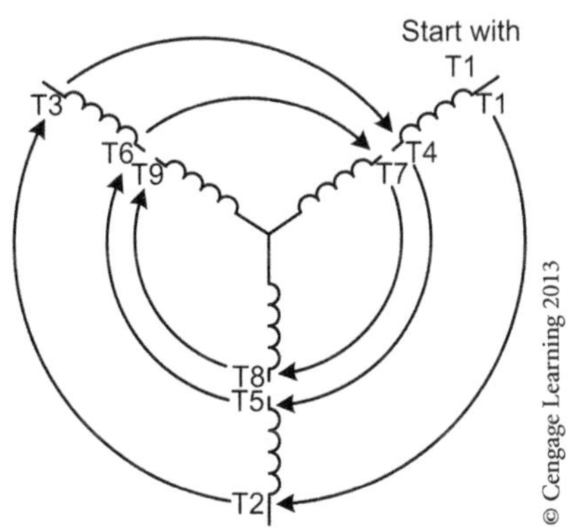

The method for labeling the 12-lead motor coils for the delta configuration connection is shown in two parts in Figure 5-6. Starting on the beginning end of any phase coil when rotating in a clockwise direction, label the starting point of the first coil T1. Turning in a clockwise direction, sequentially label the beginning end of the first phase coil of the remaining two phases, T2 and T3. Continue rotating in a clockwise direction, and move to the other end of the first phase coil you labeled T1, and label it T4. Continuing to turn in a clockwise direction, sequentially label the ends of the first coils in the remaining two phases, T5 and T6. To keep the drawing less congested, the remaining labels are shown in a separate illustration to the right of the first drawing, with the arrowed dotted lines.

Figure 5-6 Labeling the leads of a delta-connected motor

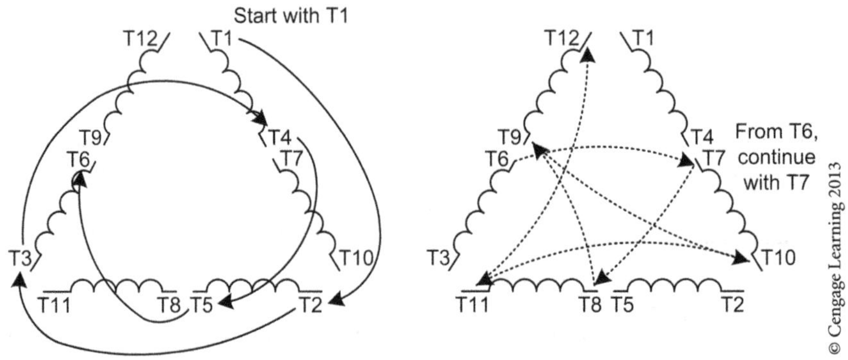

Figure 5-7 shows that the nine-lead delta motor is labeled in exactly the same way, only the coil leads T10, T11, and T12 are internally connected in the motor and are not brought out to the motor terminal box. The manufacturer internally connects coil leads T1 and T12, and brings the lead out to the motor terminal box as T1. Likewise, coil leads T2 and T10 are internally connected and brought out to the motor terminal box as T2, and coil leads T3 and T11 are connected and brought out as T3.

Figure 5-7 Labeling the leads of a nine-lead delta-connected motor

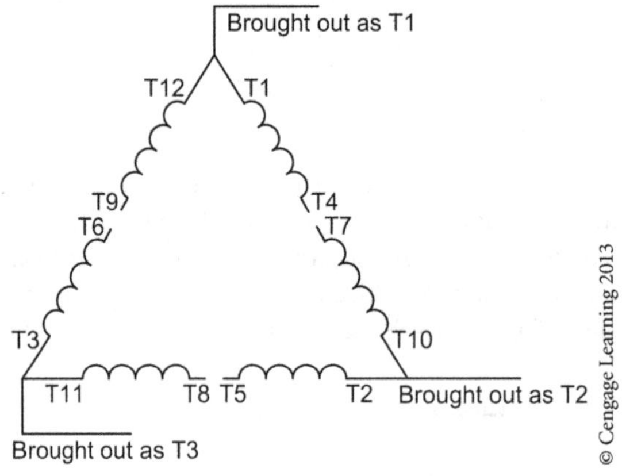

Whether a three-phase nine-lead motor is internally connected as a wye or delta, only the coil leads labeled T1 through T9 are brought out to the motor's terminal box, and coil leads T10 through T12 are not individually accessible. For the electrician, whether the motor is internally connected as a wye or delta is of little consequence, as long as the electrical supply voltage matches the rated connection voltage of the motor. The reason it is brought up here is because how the nine-lead motor is connected internally makes a big difference in the continuity measurements between the two configurations.

CONTINUITY TEST OF A NINE-LEAD WYE INTERNALLY CONNECTED THREE-PHASE MOTOR

This section of the lab is going to assume that the motor leads for all motor stator coil leads are brought out to a terminal block, similar to what is shown in Figure 5-8. If the motor you are using for this lab does not have the individual motor leads brought out to a terminal block, you will need to make appropriate changes to accommodate this lab.

Figure 5-8 Three-phase motor terminal block

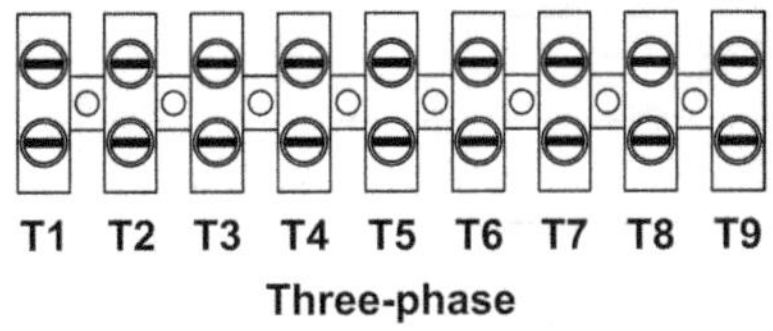

Step 6

The drawing in Figure 5-9 shows the coil configuration and internal coil connections of a three-phase nine-lead wye-configured motor. Similar to the single-phase motor, none of the T-labeled motor stator coil leads should read continuity to the motor housing. With your test meter set to continuity, complete the measurements for Table 5-5.

Figure 5-9 Nine-lead wye-connected motor

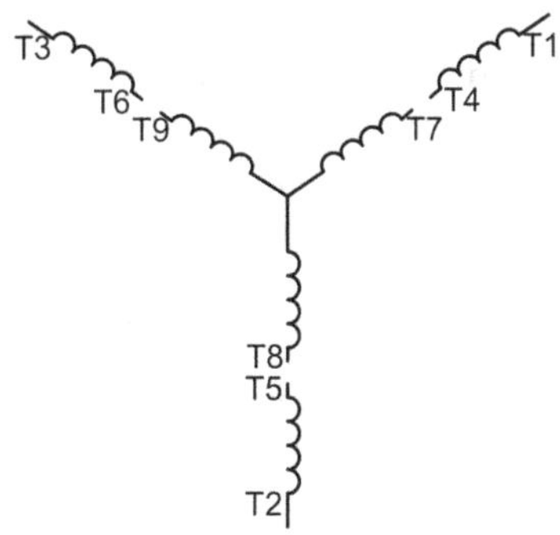

Three-Phase WYE Connected Motor Continuity to Motor Housing					
T1 to Motor Housing	T2 to Motor Housing	T3 to Motor Housing	T4 to Motor Housing	T5 to Motor Housing	T6 to Motor Housing
no	no	no	no	no	no
T7 to Motor Housing	T8 to Motor Housing	T9 to Motor Housing			
no	no	no			

Table 5-5 Three-phase wye continuity test to motor housing

1. Did any of the T-lead continuity readings in Table 5-5 read continuity to the motor housing? NO
   If any of the other readings showed continuity, the motor is damaged; do not use this motor for further testing.

Step 7

Coils T1-T4, T2-T5, and T3-T6 should all read continuity, but T1, T2, and T3 should not read continuity between them. Coil leads T7, T8, and T9 should all read continuity between them. With your test meter set to continuity, complete the measurements for Table 5-6.

Three-Phase WYE Connected Motor Continuity Test Between Coils					
T1 to T4	T2 to T5	T3 to T6	T7 to T8	T8 to T9	T9 to T7
yes	yes	yes	yes	yes	yes
T1 to T7	T2 to T8	T3 to T9			
no	no	no			

Table 5-6 Three-phase wye continuity test between coils

1. All of the continuity readings in the top row of Table 5-6 should have read continuity—did they? yes
2. All of the continuity readings in the bottom row of Table 5-6 should have read open—did they? yes
   If the answer to either questions 1 or 2 is no, the motor is damaged; do not use this motor for further testing.

Step 8

In this step you will take actual resistance readings of the motor coils. Reading the actual resistance of the individual coils does not add any validity to the continuity readings made to this point, because you have no idea what the actual resistance reading should be. The only significance of measuring actual resistance of the individual stator coils is to verify that all of the resistances are the same. With your test meter set to resistance, complete the measurements for Table 5-7.

Three-Phase WYE Connected Motor Resistance of Coils					
T1 to T4	T2 to T5	T3 to T6	T7 to T8	T8 to T9	T9 to T7
21.2 Ω	20 Ω	20.2 Ω	38.8 Ω	38.8 Ω	39 Ω

Table 5-7 Three-phase wye resistance of coils

1. The first three readings in Table 5-7 should have been the same, and the last three readings should have been the same—were they? yes

Step 9

The drawing in Figure 5-10 shows the coil configuration and internal coil connections of a three-phase nine-lead delta-configured motor. Similar to other motors tested to this point, none of the T-labeled motor stator coil leads should read continuity to the motor housing. With your test meter set to continuity, complete the measurements for Table 5-8.

Figure 5-10 Continuity test of a nine-lead delta-connected motor

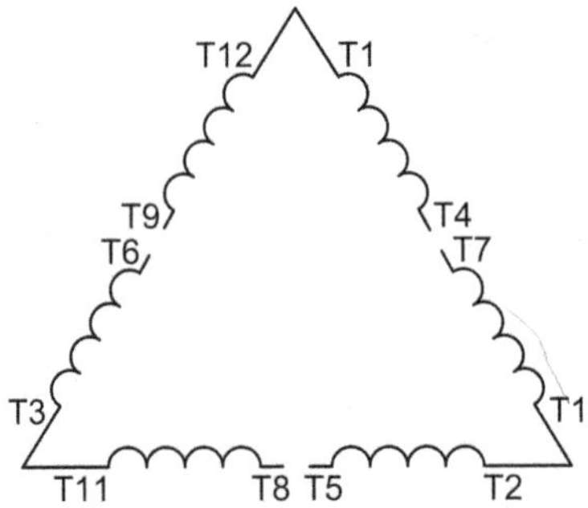

Three-Phase Delta Connected Motor Continuity to Motor Housing					
T1 to Motor Housing	T2 to Motor Housing	T3 to Motor Housing	T4 to Motor Housing	T5 to Motor Housing	T6 to Motor Housing
no	no	no	no	no	no
T7 to Motor Housing	T8 to Motor Housing	T9 to Motor Housing			
no	no	no			

Table 5-8 Three-phase delta continuity test to motor housing

Step 10

Coil lead T1 should read continuity to both T4 and T9, T2 should read continuity to both T5 and T7, and T3 should read continuity to both T6 and T8; but T1, T2, and T3 should not read continuity between them. With your test meter set to continuity, complete the measurements for Table 5-9.

Three-Phase Delta Connected Motor Continuity Test Between Coils					
T1 to T4	T1 to T9	T2 to T5	T2 to T7	T3 to T6	T3 to T8
yes	no	yes	no	yes	no
T4 to T9	T5 to T7	T6 to T8			
no	no	no			

Table 5-9 Three-phase delta continuity test between coils

1. Did all of the coil tests read continuity? no

QUESTIONS

1. What two minimum tests should be performed on motors when they arrive on the jobsite?
Continity, physical
2. Assuming that the manufacturer shipped a motor free of defects, what is the purpose of these tests?
Shipping damage
3. Why is it important to perform the test on motors before the delivery person leaves?
So you isolate them
4. When a digital multimeter reads "OL," does that mean it is measuring a short circuit or an open circuit?
open load
5. When a digital multimeter reads "0," does that mean it is measuring a short circuit or an open circuit?
Short
6. Write down the colors associated with each of the following single-phase motor terminal leads.
T1 = Blue T2 = White T5 = Black T8 = red
7. Which of the stator coil winding terminal leads should measure continuity to the motor housing?
ground or 9
8. What type of motor testing is a megger best suited for?
insulation
9. Which single-phase motor coil winding has the higher resistance, the auxiliary start winding or the main run winding?
main run
10. What are the names of the two different three-phase configurations?
wye delta
11. What are the different numbers of the stator coil leads brought out the motor terminal box for three-phase motors?
1, 2, 5, 8
12. What number of stator coil leads brought out to the motor terminal box is the most common? 4 & 9
13. What is the advantage of a motor with two stator coil windings per phase?
more torque

14. For three-phase nine-lead motors, which T leads are *not* brought out to the motor terminal box?

15. What two reasons are given in this lab for reducing the number of leads in the motor terminal box?

16. For internal-wye-connected motors, how is the T10, T11, and T12 internal connection accessed for field wiring? neutral

17. For internal-delta-connected motors, which coil leads are internally connected to the following leads that are brought out to the motor terminal box?

T1 T4 T2 T5 T3 T6

18. Number the motor coil winding leads of both the wye and delta configurations.

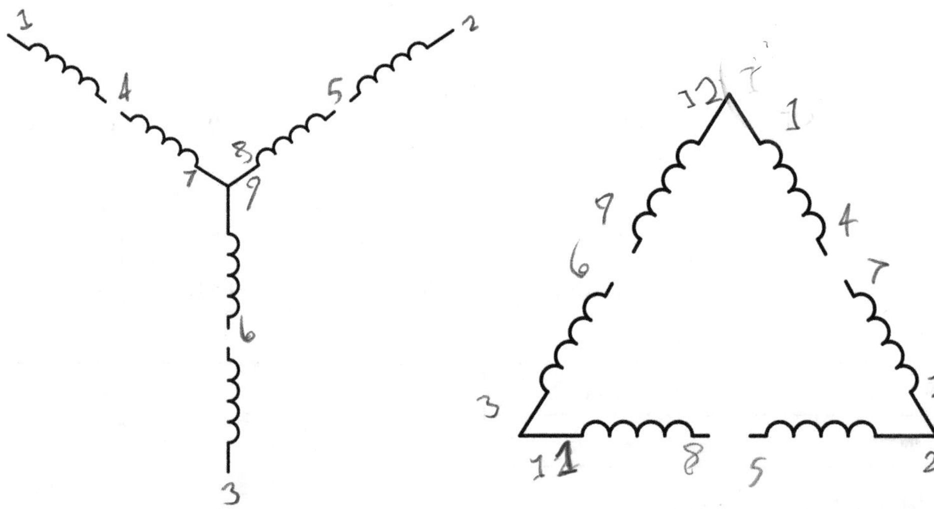

LAB EXERCISE #6

Single-Phase Motor Connections

Name:	Date:	Grade:

Purpose

The purpose of this lab is to understand the stator coil lead connections of single-phase motors, practice making the electrical connections to single-phase motors, and practice measuring electrical parameters with an electrical test meter.

Materials:

Single-voltage, reversible, single-phase 120-volt motor

Clamp meter

SINGLE-PHASE MOTOR CONFIGURATIONS

Single-phase induction motors are manufactured in a wide variety of configurations to meet the diverse needs of applications in the electrical industry. The number of stator coil connection terminal leads that are brought out to the motor's terminal box is determined by the type of induction motor it is, and the number of internal connections that are made at the factory. Motor manufacturers make these internal connections to simplify the field connection and reduce wiring errors by providing fewer coil winding leads in the terminal box, and to reduce installation time by requiring fewer terminal conductor terminations in the field.

For example, a two-terminal 115-volt single-phase motor already has the connections made between the main run winding and the auxiliary start winding, the connection to the auxiliary start switch, and the connection to the start capacitor. Only two terminal conductors are brought out to the motor's terminal box, and what could be simpler than terminating the two conductors of the branch circuit feeding the motor to the two terminal conductors in the motor? In this case, the internal connections made at the factory did, indeed, simplify the field connection of the motor.

As helpful as the internal factory connections are, each one reduces the application flexibility of the motor. Because of the internal connections made at the factory, which are not accessible in the field, the two-terminal 115-volt single-phase motor just mentioned is severely limited in its application. The main run windings and the auxiliary start windings are not individually accessible, so the motor cannot be reversed; and the motor is rated for only a single voltage.

The motor that will be referenced in this lab is a single-voltage, reversible, single-phase, 120-volt induction motor. The motor is single-voltage, so it only has one main run winding and one auxiliary start winding, but both leads of both coil windings are brought out to the motor terminal box, where they may be accessed individually. If the motor you are using for this lab is different, you will have to make some modifications to the motor connections to make it work.

This lab is going to assume that the motor leads for the auxiliary start and main run windings of the motor you are using for the lab are brought out to a terminal block, similar to what is shown in Figure 6-1. If the motor you are using for this lab does not have the individual motor leads brought out to a terminal block, Table 6-1 identifies the lead conductor colors for you to locate in the motor terminal box.

Figure 6-1 Single-phase motor terminal block

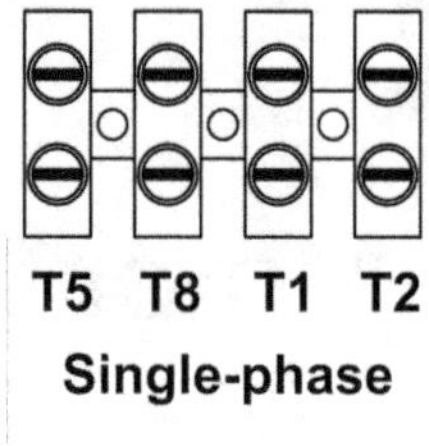

NEMA Single-Phase Lead Colors
T1–Blue
T2–White
T5–Black
T8–Red

Table 6-1 NEMA single-phase motor lead colors

Step 1

In this step the single-phase 120-volt power source will be verified. The power source for this lab will be shown in the diagrams as a terminal block; it may be necessary for you to make adjustments if the power source you are using is a different configuration. Energize the power source to verify the voltage level is 120 volts, and record your reading in Table 6-2.

Verify 120-V Source
Line to Nuetral Voltage Reading of Power Source
120 V

Table 6-2 Verify 120-V source

Step 2

In this step you will connect the single-phase motor, as shown in Figure 6-2, to measure the motor's running current and the motor's direction of rotation. To measure the motor current, clamp your clamp meter around the T1 lead of the single-phase motor. You may include the auxiliary start winding T5 lead also, but it will not make any difference in your reading; the centrifugal start switch will take the auxiliary start winding out of the circuit immediately after the motor rotor starts turning. Take the current readings and document the direction of rotation to complete Table 6-3 after the motor is running.

Figure 6-2 Current readings of single-phase motor

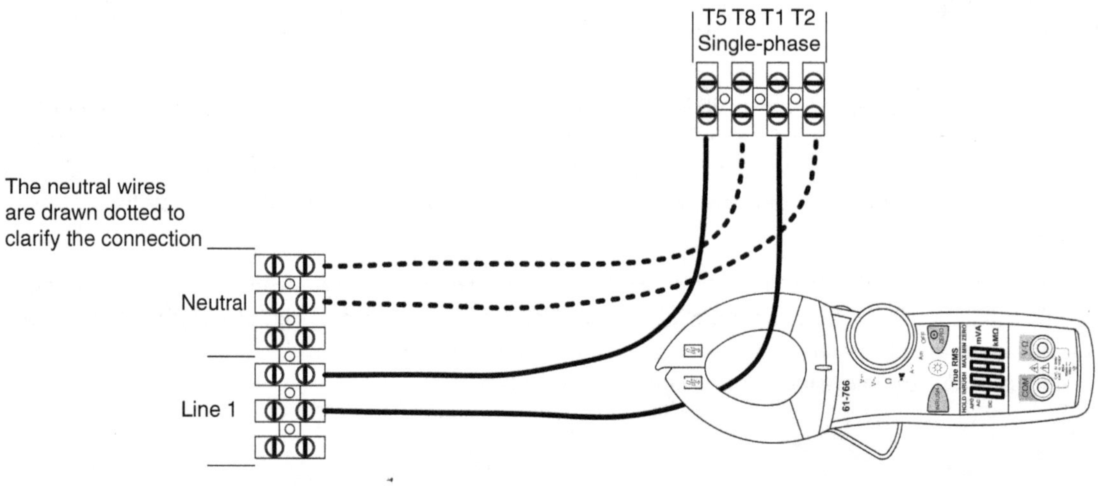

Current Readings of Running Single Phase Motor, and Rotation		
Current Reading of Motor Terminal T5	Current Reading of Motor Terminal T1	Direction of Rotation from Non-Drive End
0 A	5.64 A	Clockwise

Table 6-3 Current readings of running single-phase motor and rotation

1. Why is the current reading of T5 zero amperes?

Its the start winding after disconnection

Step 3

In this step you will reverse the direction of rotation that the motor turns, from Table 6-3 of Step 2. Figure 6-3 shows the instantaneous magnetic pole polarity of the auxiliary start winding in relation to the instantaneous magnetic pole polarity of the main run winding. Without getting into which magnetic pole relationships cause which directions of rotation, it is sufficient to understand that if having two like-magnetic-polarity poles next to each other causes one direction of rotation, having two opposite-magnetic-polarity poles next to each other will cause the opposite direction of rotation.

Figure 6-3 Reversing a single-phase motor graphic

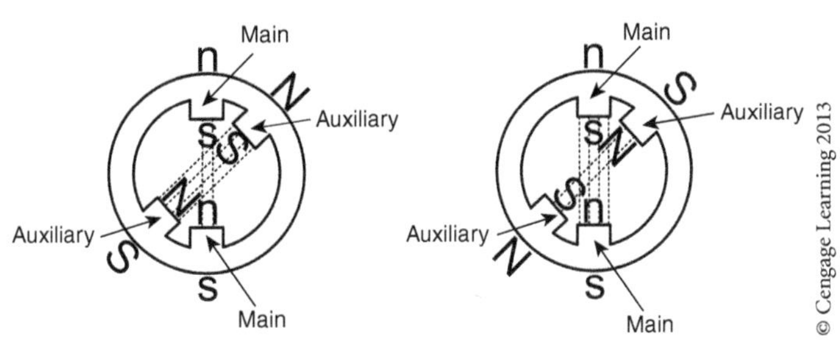

Modify the wiring of the circuit shown in Figure 6-2 in Step 2 to interchange the leads of the auxiliary start winding, T5 and T8, as shown in Figure 6-4. Take the current readings and document the direction of rotation to complete Table 6-4 after the motor is running.

Figure 6-4 Interchange auxiliary start winding leads

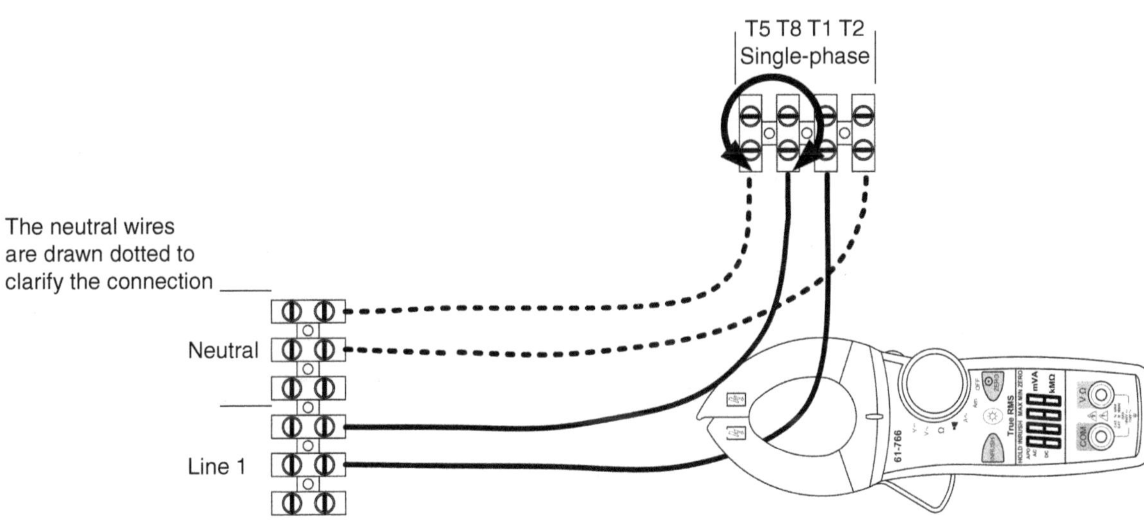

Current Readings of Running Single Phase Motor, and Rotation	
Current Reading of Motor Terminal T1	Direction of Rotation from Non-Drive End
5.63 A	Counter clockwise

Table 6-4 Current readings of running single-phase motor and rotation

1. Did the motor's direction of rotation change when the auxiliary start winding leads were interchanged in relation to the main run winding leads (from Table 6-3 to Table 6-4)? Yes
2. Did the motor's running current change when the direction of rotation was changed (from Table 6-3 to Table 6-4)?

NO

Step 4

In this step you will measure the motor's locked-rotor current. There are actually two types of locked rotor that a motor may experience: locked rotor because of a mechanical overload condition that prevents the motor shaft from turning, or locked rotor because the auxiliary start winding does not energize for some reason. Mechanical-overload locked rotor means that the pull-up torque of the motor is less than the torque required to turn the mechanical load, so the motor stalls and experiences a locked-rotor condition. In this condition, both the auxiliary start winding and the main run winding will draw current, which means the locked-rotor current will be much higher. Trying to simulate a mechanical-overload locked-rotor condition by preventing the motor rotor from turning, however, can be dangerous. For that reason the locked-rotor condition simulated here will be accomplished by disconnecting the auxiliary start winding and energizing only the main run winding.

To measure the locked-rotor current of the motor, take off either the T5 lead or the T8 lead, or both. When the motor is started, the auxiliary start winding will not be connected so the rotor will not have any incentive to turn, and the main run winding will draw its locked-rotor current. The only drawback to measuring locked-rotor current with this method is that the auxiliary start winding will not draw any current because it is disconnected, so the current reading will not be the true motor locked-rotor current. The purpose of this measurement is not to replicate actual motor nameplate locked-rotor current; rather, it is simply to observe the large increase in the main run winding current at locked rotor, compared to its normal running current. This is still a helpful exercise for the learner to experience, because most of the problems associated with single-phase motors not starting are caused by auxiliary start winding circuit problems, a malfunctioning centrifugal weight assembly, a malfunctioning start switch contact, or a bad starting capacitor.

When the motor is energized in this step, the high current draw can damage the motor stator windings in a short time. DO NOT LEAVE THE MOTOR ENERGIZED ANY LONGER THAN ABSOLUTELY NECESSARY TO TAKE THE CURRENT READING—ABSOLUTELY NO LONGER THAN 3 SECONDS! YOU DO NOT NEED TO ALLOW THE METER TO COMPLETELY STABILIZE; A ROUGH MEASUREMENT IS SUFFICIENT FOR THIS STEP. DO NOT PERFORM THIS STEP MORE THAN ONCE, BECAUSE THE MOTOR STATOR WINDINGS WILL NOT HAVE SUFFICIENT TIME TO DISSIPATE THE ADDITIONAL HEAT CAUSED BY LOCKED-ROTOR CURRENT BETWEEN TESTS, AND DAMAGE WOULD LIKELY OCCURE WITH MULTIPLE TESTS. Energize the motor and measure the locked-rotor current of the main run winding, and record your reading in Table 6-5.

Locked Rotor Current
Current Reading of Motor Terminal T1
18 A

Table 6-5 Locked-rotor current

1. Compare the running current readings of the motor's main run winding coil in Table 6-4 with the locked-rotor current readings of the same winding coil in Table 6-5. Which is higher, running current or locked-rotor current? locked Rotor
2. Relating your answer to the speed of cutting action, and the fact that the motor rotor is standing still, why is the motor's locked-rotor current higher than the running current?

magnetic field changing & resisting changing polarity

Step 5

In this step you will observe the operation of the auxiliary start winding switch: closing when the motor approaches a stopped condition, and opening when the motor approaches operating speed. Connect the circuit shown in Figure 6-5. Close the toggle switch and energize the power source; the motor will start normally. With the motor running at operating speed, the start winding is de-energized by the start switch and centrifugal weight assembly. Open the toggle switch, which will de-energize the main run winding, and the motor will coast toward a stop.

Before the motor completely stops, however, you will hear the centrifugal weight assembly close the start switch and re-energize the start winding. Because the motor has not yet stopped turning, the start winding alone will be able to bring the motor back up close to operating speed from that remaining rotation, where the centrifugal weight assembly and start switch will again de-energize the start winding and the motor will coast toward a stop. Understand what to expect, and be ready to make your observations before energizing this circuit. Do not allow this process to repeat itself more than three or four times, or the auxiliary start windings may build up excessive heat.

Figure 6-5 Start switch observation

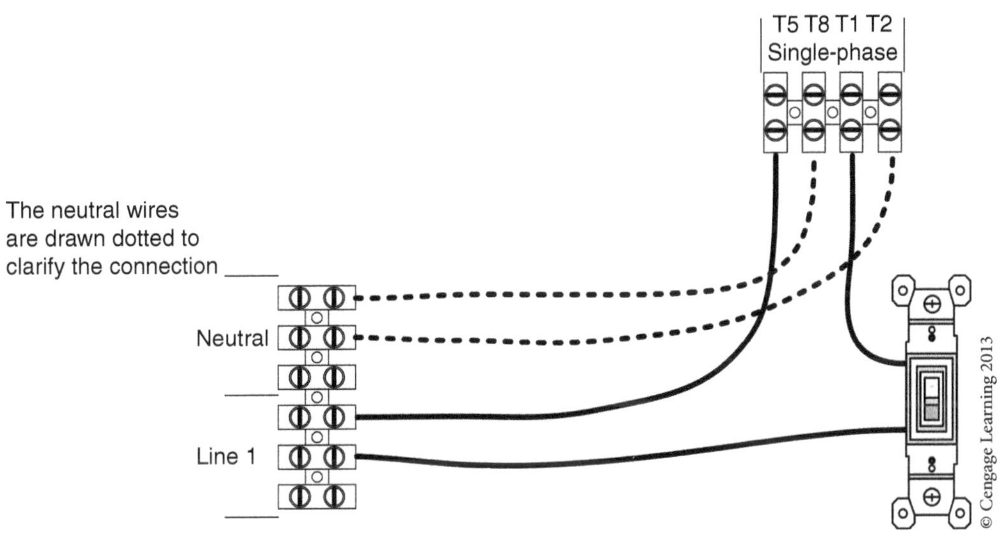

QUESTIONS

1. Why are motors with internal coil connections less versatile in applications than motors with all coil connection leads brought out to the motor terminal box?
   Can't change them
2. At what part of the motor's running cycle is the auxiliary start winding de-energized?
   When its at full speed
3. At what part of the motor's running cycle is the auxiliary start winding re-energized?
   when the rotor is stopped
4. The direction of rotation for a single-phase motor is determined by the instantaneous magnetic pole polarities between what? Start / run coil
5. How is the direction of rotation reversed for a single-phase motor?
   reversing start winding
6. What are two types of locked-rotor conditions that single-phase motors may experience?
   Physical overload & no start winding
7. Which locked-rotor condition will cause a greater current draw from the power supply?
   Physical overload
8. What T labels are given to the connection lead conductors of the auxiliary start winding?
   T5 & T8
9. What T labels are given to the connection lead conductors of the main run winding?
   T1 & T2

LAB EXERCISE #7

Three-Phase Motor Connections Lab

Name:	Date:	Grade:

Purpose

The purpose of this lab is to understand the stator coil lead connections of three-phase motors, practice making the electrical connections to three-phase nine-lead motors, and practice measuring electrical parameters with an electrical test meter.

Materials:

Three-phase nine-lead motor

Clamp meter

THREE-PHASE MOTOR CONFIGURATIONS

Three-phase induction motors are manufactured in a wide variety of configurations to meet the diverse needs of applications in the electrical industry. The number of stator coil connection terminal leads that are brought out to the motor's terminal box is determined by the number of stator coil windings the motor has, and the number of internal connections that are made at the factory. Motor manufacturers make these internal connections to simplify the field connection and reduce wiring errors by providing fewer coil winding leads in the terminal box, and to reduce installation time by requiring fewer terminal conductor terminations in the field.

Nine-lead three-phase motors are the most common, because they represent a good balance between internal coil winding connections, while still retaining the flexibility of having a dual-voltage rating. The dual-voltage rating is important, because of the diverse three-phase electrical system voltages available in buildings. For instance, a small commercial building may have a lower 120/240-volt or 120/208-volt service, whereas an industrial building in the same area may have a higher 240/480-volt or 277/480-volt service. By manufacturing a motor with two stator coil windings per phase, the same motor may be installed in either of the lower-voltage systems (208 or 240 volts) or higher-voltage systems (480 volts).

A common dual-voltage rating for motors manufactured to be used in these voltage level systems is 230/460 volts. Even though the 208-volt voltage level of the 120/208-volt wye electrical system does not match the 230-volt low-voltage rating of the motor, it is a common practice to install these motors on both the 208-volt and 240-volt systems interchangeably. In more recent years, many motor manufacturers have modified their voltage ratings of these motors to 208–230/460 volts, and even 200–230/480 volts, to accommodate these voltage discrepancies.

The dual-voltage rating of these motors is attained by manufacturing the motor with two stator coils per phase, and then connecting the coils in parallel for the low-voltage connection, or in series for the high-voltage connection. When the phase coil windings are connected parallel, the full-phase voltage will be applied to each of the phase coil's windings. When the phase coil windings are connected in series, however, only half of the full-phase voltage will be applied to each of the phase coil windings. With the phase coils connected in series, twice as much phase voltage may be applied to the motor.

HIGH-VOLTAGE WYE CONNECTION

The drawing in Figure 7-1 shows the high-voltage coil connections for a wye-connected motor. Notice that the power supply line conductors L1, L2, and L3 connect only to motor stator coil winding leads T1, T2, and T3, respectively; th line conductors do not connect to any other coil winding leads. Motor coil winding terminals T10, T11, and T12 are not shown on the connection diagram, because they are already connected internally, and cannot be accessed.

Figure 7-1 NEMA nine-lead high-voltage wye-connected motor

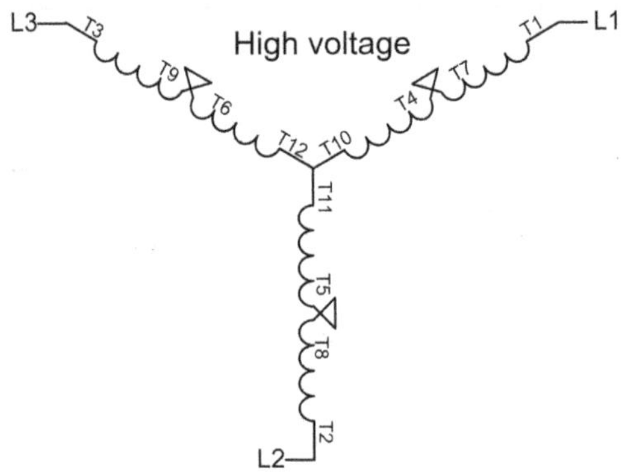

VOLTAGE	L1	L2	L3	JOIN
HIGH	1	2	3	4,7
				5,8
				6,9

LOW-VOLTAGE WYE CONNECTION

The drawing in Figure 7-2 shows the low-voltage coil connections for the wye-connected motor. Notice that the power supply line conductors L1, L2, and L3 each connect to two motor stator coil winding leads. Power supply line conductor L1 connects to both stator coil winding leads T1 and T7; L2 connects to T2 and T6; and L3 connects to T3 and T9. In order to connect the two stator coil windings of each phase in parallel with each other, T4, T5, and T6 would have to connect to the same common point as T10, T11, and T12; but because the T10, T11, and T12 common point is not accessible, T4, T5, and T6 are simply connected to form a second wye connection in the motor. Sometimes the low-voltage wye connection is documented on motor nameplates and connection diagrams as YY, possibly to signify the double-wye configuration.

Figure 7-2 NEMA nine-lead low-voltage wye-connected motor

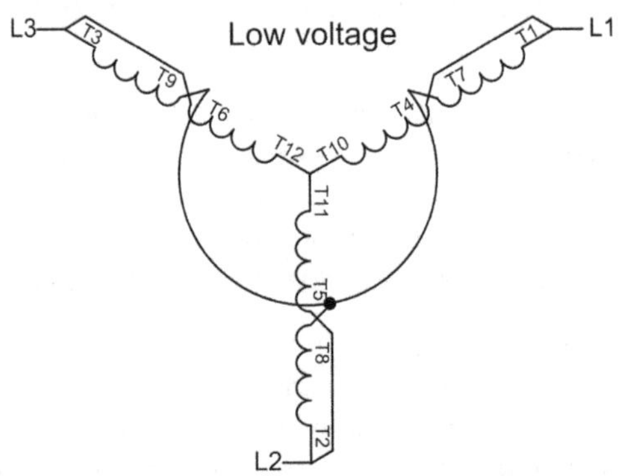

VOLTAGE	L1	L2	L3	JOIN
LOW	1,7	2,8	3,9	4,5,6

HIGH-VOLTAGE DELTA CONNECTION

The drawing in Figure 7-3 shows the high-voltage coil connections for the delta-connected motor. Notice that even though it appears that each of the power supply line conductors connects to two motor stator coil winding leads, the conductors are not connected in this way. In the drawing it appears that L1 is connected with stator coil winding leads T1 and T12, but remember that coil winding leads T1 and T12 are connected internally in the motor and brought out to the motor terminal box as T1. The same is true for T2 and T10, and T3 and T11. Look at the high-voltage connection diagram for the wye-configuration motor, and notice that it is identical to the high-voltage connection diagram for the delta-configuration motor.

Figure 7-3 NEMA nine-lead high-voltage delta-connected motor

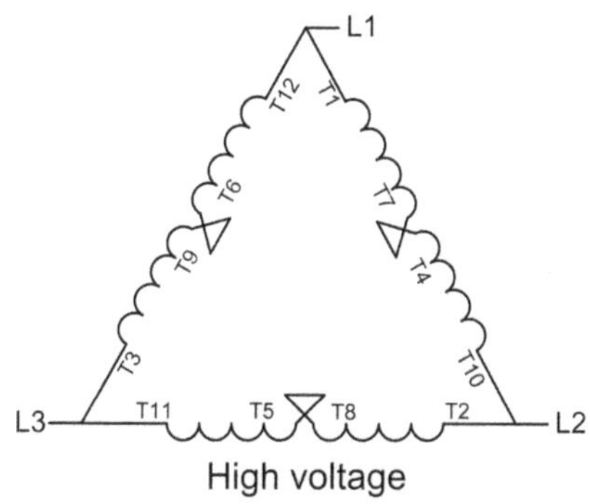

VOLTAGE	L1	L2	L3	JOIN
HIGH	1	2	3	4,7
				5,8
				6,9

LOW-VOLTAGE DELTA CONNECTION

The drawing in Figure 7-4 shows the low-voltage coil connections for the delta-connected motor. Notice that the power supply line conductors L1, L2, and L3 each connect to three motor stator coil winding leads. Power supply line conductor L1 connects to stator coil winding leads T1, T6, and T7; L2 connects to T2, T4, and T8; and L3 connects to T3, T5, and T9. Sometimes the low-voltage delta connection is documented on motor nameplates and connection diagrams as a ΔΔ configuration.

Figure 7-4 NEMA nine-lead low-voltage delta-connected motor

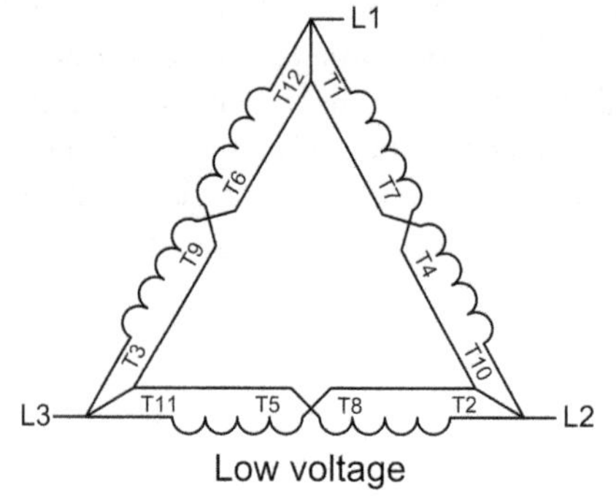

VOLTAGE	L1	L2	L3	JOIN
LOW	1,6,7	2,4,8	3,5,9	--

REVERSING THE DIRECTION OF ROTATION

The three-phase motor works on the principle that the magnetic polarity poles of the three individual sets of electromagnet coil windings placed at 120° intervals around the motor stator will rise and fall in sequence to create a rotating magnetic field. Figure 7-5 shows a drawing with the three phases A, B, and C placed 120° out-of-phase with each other to demonstrate this point. The three phases can only follow one of two different sequences—either ABC, or ACB; there are no other possibilities. In the drawing, sequence ABC would be a clockwise direction of rotation, and sequence ACB would be a counterclockwise direction of rotation. A three-phase motor has a 50/50 chance that it will turn in the desired direction of rotation the first time it is energized, and may be reversed by exchanging any two power supply line leads.

Figure 7-5 Pole relationships for clockwise (CW) or counterclockwise (CCW) rotation diagram

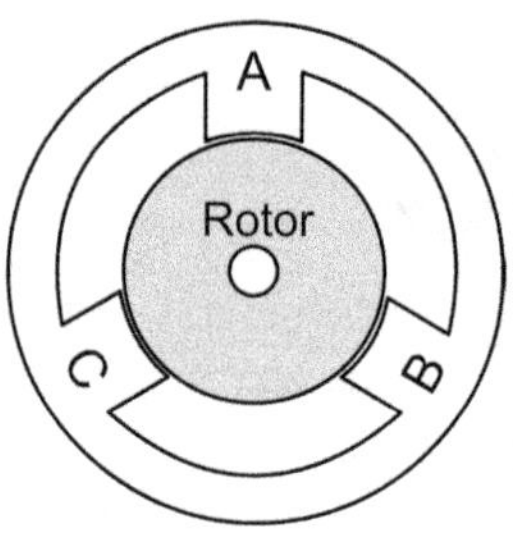

This lab is going to assume that a nine-lead three-phase motor is being used, and that the motor leads are brought out to a terminal block, similar to what is shown in Figure 7-6. If the motor you are using for this lab does not have the individual motor leads brought out to a terminal block, the connection leads in the motor terminal box should be labeled with the T numbers 1 through 9.

Figure 7-6 Three-phase motor terminal block

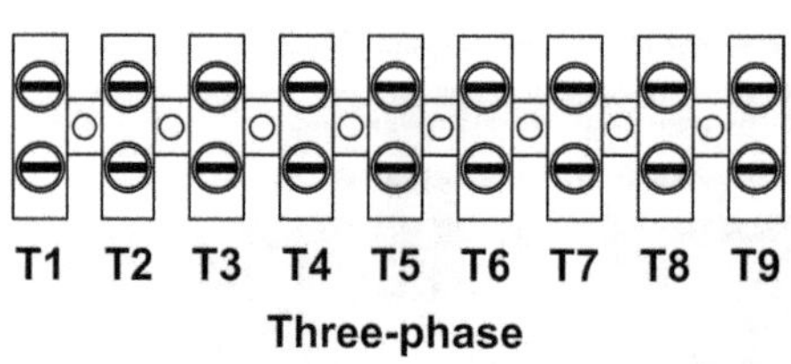

Step 1

In this step you will connect the three-phase wye internally connected motor for the high-voltage connection to a three-phase 480-volt power source, and measure running current. Use the circuit drawing and high-voltage connection table shown in Figure 7-7. Use your test clamp meter set to current, and complete the measurements for Table 7-1.

Figure 7-7 480-volt wye connection to power source

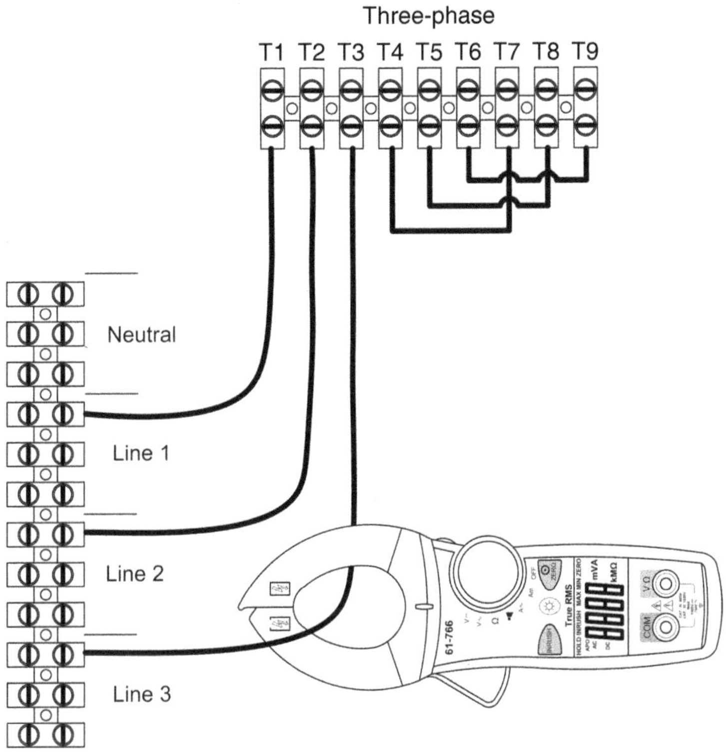

VOLTAGE	L1	L2	L3	JOIN
HIGH	1	2	3	4,7
				5,8
				6,9

Connection table

Three-Phase Wye 480-V Current Readings		
T1	T2	T3
660 mA	630 mA	640 mA

Table 7-1 Three-phase wye 480-V current readings

1. Are all of the current readings about equal? yes
2. Why doesn't the motor draw the current listed on the nameplate?

   not full load

Step 2

If you are using a fractional-horsepower three-phase motor for this lab, your current reading is very small, probably less than 1 ampere. In this step you will loop the conductor you are measuring the current of through the meter clamp multiple times, as shown in Figure 7-8, to get a more accurate reading. Wire nut conductors together, if necessary, until the length of the conductor is sufficient to have 10 turns of wire through the meter clamp. Take the current readings of all three phases again, using the 10-turn multiplier loop. Then move the decimal point one place to the left to calculate the actual current reading.

Figure 7-8 Ten-turn coil current reading

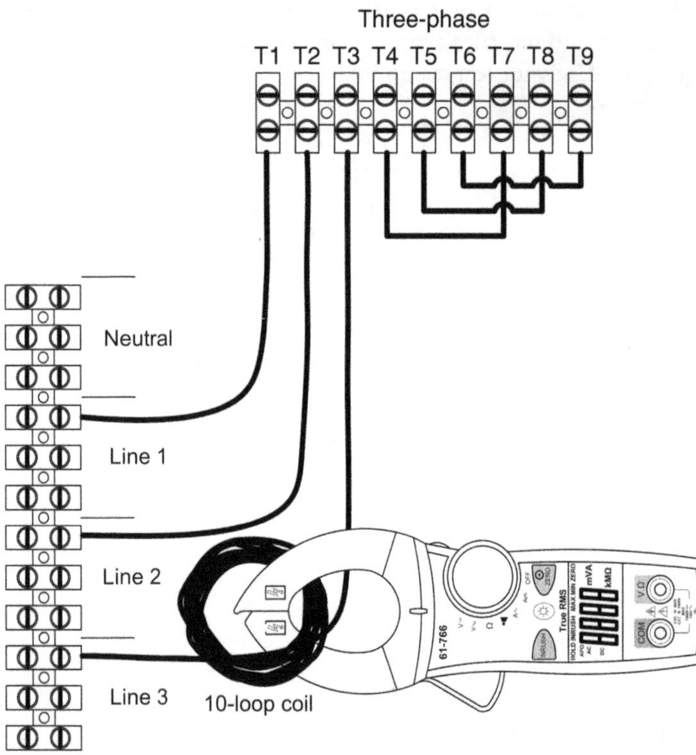

Three-Phase Wye 480-V Current Readings with a 10-Turn Coil		
T1	T2	T3
6.83 A	6.68 A	6.83 A
Divided by 10 =	Divided by 10 =	Divided by 10 =
683 mA	668 mA	683 mA

Table 7-2 Three-phase wye 480-V current readings with 10-turn coil

1. After dividing the measured line current using the 10-turn coil in Table 7-2, were your measurements the same as your measurements without the 10-turn coil measured in Table 7-1?

   yes

As electricians, we normally do not require such accurate readings, but it is good to know and be able to apply this method of reading very low currents with clamp-on ammeters of lower resolution.

Step 3: Locked-rotor current

To measure the locked-rotor current of the motor, remove any one of the three line leads from the power source. When the motor is started, there will only be two of the three phases connected to the motor, so the rotor will not have any incentive to turn, and the motor will draw its locked-rotor current. When you energize the motor in this step, the high current draw can damage the motor stator windings in a short time. DO NOT LEAVE THE MOTOR ENERGIZED ANY LONGER THAN ABSOLUTELY NECESSARY TO TAKE THE CURRENT READING—ABSOLUTELY NO LONGER THAN 3 SECONDS! YOU DO NOT NEED TO ALLOW THE METER TO COMPLETELY STABILIZE; A ROUGH MEASUREMENT IS SUFFICIENT FOR THIS STEP. DO NOT PERFORM THIS STEP MORE THAN ONCE, BECAUSE THE MOTOR STATOR WINDINGS WILL NOT HAVE SUFFICIENT TIME TO DISSIPATE THE ADDITIONAL HEAT, AND DAMAGE WOULD LIKELY OCCUR WITH MULTIPLE TESTS. Energize the motor and take the locked-rotor current on one of the two remaining energized lines. Record your current reading in Table 7-3.

Locked Rotor Current
Current Reading of One Energized Motor Terminal
2.74 A

Table 7-3 Locked-rotor current

1. How many times higher was the locked-rotor current of this motor as compared to the rated running current full-load amperes?

Step 4

In this step you will connect the three-phase wye internally connected motor for the low-voltage connection to a three-phase 208-volt power source, and measure running current. If the only three-phase power source you have available is 240 volt, it will still work satisfactorily for this step in the lab with an unloaded motor. Use the circuit drawing and connection table shown in Figure 7-9. Use your test clamp meter set to current, and complete the measurements for Table 7-4.

Figure 7-9 208-volt wye connection to power source

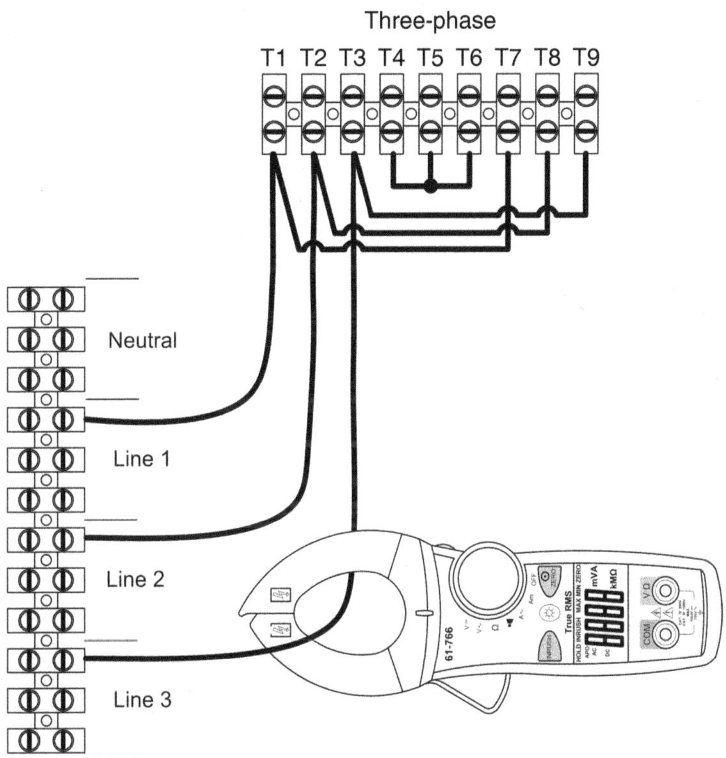

VOLTAGE	L1	L2	L3	JOIN
LOW	1,7	2,8	3,9	4,5,6

Connection table

Three-Phase Wye 208-V Current Readings		
T1	T2	T3
990 mA	950 mA	920 mA

Table 7-4 Three-phase wye 208-V current readings

1. Were all of the current readings equal?

Yes

Step 5

In this step you will connect the three-phase delta internally connected motor for the high-voltage connection to a three-phase 480-volt power source, and measure running current. Use the circuit drawing and high-voltage connection table shown in Figure 7-10. Use your test clamp meter set to current, and complete the measurements for Table 7-5.

Figure 7-10 480-volt delta connection to power source

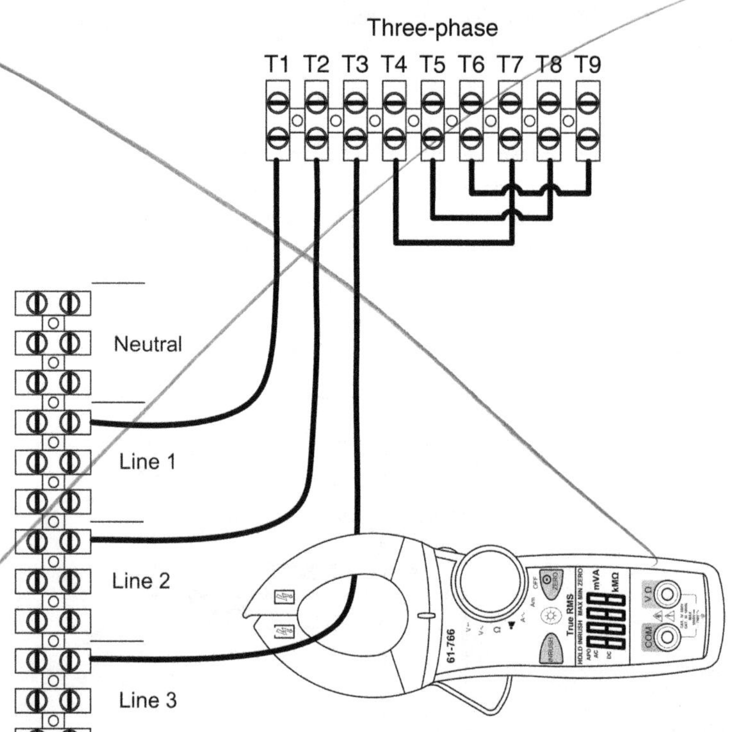

VOLTAGE	L1	L2	L3	JOIN
HIGH	1	2	3	4,7
				5,8
				6,9

Connection table

Three-Phase Delta 480-V Current Readings		
T1	T2	T3

Table 7-5 Three-phase delta 480-V current readings

1. Were all of the current readings equal?

Step 6

In this step you will connect the three-phase delta internally connected motor for the low-voltage connection to a three-phase 240-volt power source, and measure running current. If the only three-phase power source you have available is 208 volts, it will still work satisfactorily for this step in the lab with an unloaded motor. Use the circuit drawing and high-voltage connection table shown in Figure 7-11. Use your test clamp meter set to current, and complete the measurements for Table 7-6.

Figure 7-11 240-volt delta connection to power source

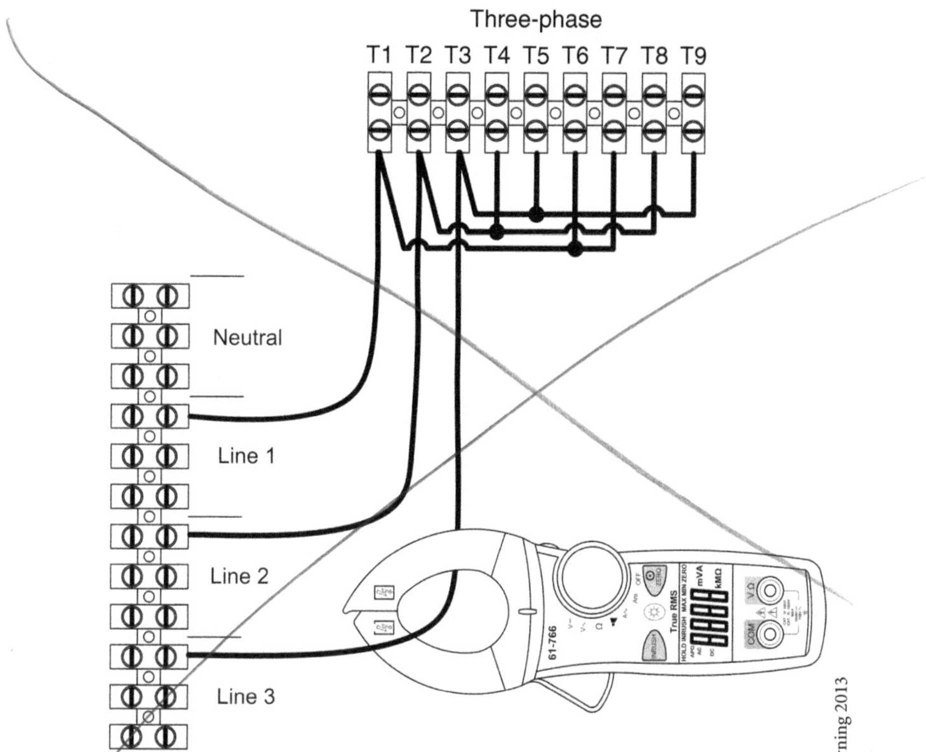

VOLTAGE	L1	L2	L3	JOIN
LOW	1,6,7	2,4,8	3,5,9	--

Connection table

Three-Phase Delta 240-V Current Readings		
T1	T2	T3

Table 7-6 Three-phase delta 240-V current readings

1. Were all of the current readings equal?

QUESTIONS

1. Why are nine-lead motors the most common three-phase motor configuration?
2. How do three-phase motors attain their dual-voltage capability?
3. Are the two stator phase coils connected in series or parallel for low-voltage operation?
4. Are the two stator phase coils connected in series or parallel for high-voltage operation?
5. On the low-voltage wye connection, how is the T10, T11, and T12 common point accessed for paralleling the phase coils?
6. What nomenclature is sometimes used on motor nameplates and connection diagrams to denote the double-wye connection?
7. What are the chances a three-phase motor will turn in the desired direction of rotation the first time it is energized?
8. How is the direction of rotation of a three-phase motor reversed?

LAB EXERCISE #8

Motor Nameplate Lab

Name:	Date:	Grade:

Purpose

The purpose of this lab is to gain experience in reading, interpreting, and understanding the induction motor rating information listed on the motor nameplate.

Material:

Nameplate information from three different induction motors

MOTOR NAMEPLATE INFORMATION

The National Electrical Manufacturers Association (NEMA) defines basic motor design and dimension parameters for medium single-phase and poly-phase induction motors in NEMA Standard MG-1. The MG-1 standard also identifies a minimum set of these parameters that must be listed on the motor nameplate. The intent of the motor nameplate information requirement is to provide a common rating system between motor manufacturers to help with interchangeability and application of the motors. It is important for the electrician to understand how to read and interpret these parameters, because they define the basic performance and mounting characteristics of the motor.

Listed here are 20 different motor parameters that may be documented on the motor nameplate, with their accompanying abbreviations for reference. This is not a complete list of all motor parameters that may be documented on the nameplate, and not all of these motor parameters are required to be documented on the nameplate, but the list represents the most commonly documented motor parameters.

1. Manufacturer's name and address
2. Rated voltage (V, or VOLT)
3. Rated full-load amperes (FLA)
4. Rated frequency (F, Hz, or FREQ)
5. Number of phases (P, or PHASE)
6. Rated full-load speed (RPM)
7. Rated horsepower (HP)
8. Locked-rotor kVA code letter (CODE)
9. Rated duty or time rating (DUTY)
10. Insulation class temperature (INS)
11. Ambient temperature (AMB or TEMP)
12. Type (TYPE)
13. Service factor (SF)
14. Frame (FR, or FRAME)
15. Design code Letter (CODE)
16. Enclosure type (ENCL)
17. Bearings (BRGS)
18. Model (MODEL)
19. Serial number (SN)
20. Connection diagrams

For this lab you will survey and document the nameplate information from three different induction motors, and then complete the analysis portion of the lab for the three motors. The motors that you survey to gather the nameplate information may not provide all 20 of the parameters just listed, or they may document other parameters that are not listed here. Document as many of the parameters listed in each survey form (see Figure 8-1) as possible for each motor.

Figure 8-1 Nameplate survey form

Motor # 1 nameplate survey

Manufacturer's Name and Address	
Phases (P, or PHASE)	Frequency (F, Hz, or FREQ)
Voltage (V, or VOLT)	Amperes (A or FLA)
Full load speed (RPM)	Horsepower (HP)
Locked rotor kVA code letter (CODE)	Duty or time rating (DUTY)
Insulation class temperature (INS)	Ambient temperature (AMB or TEMP)
Type (TYPE)	Service factor (SF)
Frame (FR, or FRAME)	Design code letter (CODE)
Enclosure type (ENCL)	Bearings (BRGS)
Model (MODEL)	Serial number (SN)
Connection diagrams	

PHASE ANALYSIS

The phase rating of the motor identifies the type of power source the motor is designed to operate on—whether the motor requires a single-phase or three-phase power source. Many years ago two-phase motors were available, but they have been obsolete for several decades and the chance that you will encounter one is very remote. Write down the number of phases each of the motors you surveyed is rated for.

Motor #1 1 Motor #2 3 Motor #3 3

FREQUENCY ANALYSIS

The frequency rating of the motor identifies the frequency, or Hz, the motor is designed to operate on. All of North America (governed by NEMA code) is standardized on 60 Hz, but other parts of the world (governed by the International Electric Code [IEC]) are standardized on 50 Hz. Motors are only designed to tolerate a frequency variation of 5 percent, but some motors are rated to operate on either 50 Hz or 60 Hz. Write down what frequency each of the motors you surveyed is designed to operate on.

Motor #1 60 Hz Motor #2 60 Hz Motor #3 60 Hz

VOLTAGE ANALYSIS

The voltage rating of the motor determines what type of electrical distribution system configuration the motor was designed to operate in. Single-phase motors that are rated 110 V, 115 V, or 120 V are designed to operate on a common 120-volt line-to-neutral branch circuit; the distribution system configuration is irrelevant.

Single-phase motors that are rated for 208 volts are designed for line-to-line operation from two of the three phases of a three-phase wye-configured electrical distribution system, which is 208 volts. Single-phase motors that are rated for 230 volts are designed for line-to-line operation from two of the three phases of a three-phase delta-configured electrical distribution system, which is 240 volts.

Single-phase dual-voltage motors are often rated 115/230 volts, which would indicate two line leads of a 240-volt delta system, but these motors are commonly connected to 208 volts for the high-voltage connection. Motors with voltage ratings of 115/208–230 volts are becoming more common, to indicate that they will perform equally well on two line leads of either a wye or delta distribution system.

Three-phase motors are determined in the same way, except that there is no three-phase line-to-neutral motor connection. All three-phase motor connections are line-to-line only. Three-phase motors that are rated for 208 volts are designed for line-to-line operation on all three line phases of a three-phase 120/208-volt wye distribution system. Three-phase motors that are rated for 230 volts are designed for line-to-line operation on all three line phases of a three-phase 120/240-volt delta distribution system.

Three-phase motors that are rated 460 volts are designed to operate line to line on all three phases of either a 277/480-volt wye-configured system or a 240/480-volt delta-configured system; the distribution system configuration is irrelevant. Determine which electrical distribution system each of the motors you surveyed is designed to operate on.

Motor #1 115 V Motor #2 115/230 V Motor #3 230Y/460Y V

AMPERE ANALYSIS

The full-load ampere (FLA) measurement of a motor is taken when the motor is loaded to the full-load horsepower using the motor's rated frequency and voltage. The motor will draw less than full-load amperes when it is less than fully loaded, and more than full-load amperes when it is overloaded.

The nameplate should list a separate current, or full-load ampere (FLA), rating for each of the voltages that the motor is rated for. For example, a motor rated for 208/230/480 volts would have three separate respective current ratings, also separated by the "/" delineator. If any of the motors you chose to survey have multiple voltage ratings, notice that the voltage and current ratings are inversely proportional; the higher voltage ratings have lower current ratings associated with them. Document the current ratings of the motors you surveyed.

Motor #1 6 A Motor #2 3.2 /1.6 A Motor #3 3.4 /1.7 A

FULL-LOAD SPEED ANALYSIS

The full-load speed of an induction motor is the synchronous speed minus the slip when the motor is operating at its rated horsepower. In most circumstances the actual full-load speed of the motor is less important than the synchronous speed the motor is designed for. The design differences between different motor manufacturers may cause one manufacturer's motor to be rated for 1,750 RPM, whereas the equivalent motor of another manufacturer is rated for 1,760 RPM. More than likely the 10 RPM is not going to make a significant difference in the operation of the load on the motor.

Synchronous speed is determined by the line frequency of the source voltage and the number of stator magnetic pole coil windings the motor has. The formula for synchronous speed is Synchronous speed = Hz * 60 (seconds in 1 minute) / Number of pairs of poles. A two-pole induction motor would have one pair of poles, and the synchronous speed would be calculated as Synchronous speed = 60 Hz*60 seconds/1 pairs of poles = 3,600/1, or 3,600 RPM. A four-pole motor would be 3,600/2 = 1,800 RPM, a six-pole motor would be 3,600/3 = 1,200 RPM, and so on.

To determine the synchronous speed of an induction motor, the number of poles is normally not listed on the motor nameplate, so it must be calculated. For the most common designs of induction motors, the slip at full load is approximately 3 percent. In the case of the 1,750 full-load RPM motor just mentioned, multiplying 1,750 RPM times 1.03 equals 1,803 RPM, and multiplying 1,760 RPM times 1.03 equals 1,813 RPM. The synchronous speeds for a 60-Hz power supply were listed earlier as 3,600 RPM, 1,800 RPM, and 1,200 RPM. It is apparent that both of these calculations are for a synchronous speed of 1,800 RPM, which is the synchronous speed of a four-pole induction motor.

Using a 3 percent slip calculation as noted earlier, calculate the synchronous speed for each of the motors you surveyed.

Motor #1 1725 RPMS Motor #2 1725 RPMS Motor #3 1750 RPMS

HORSEPOWER ANALYSIS

There really is not any analysis of the motor's horsepower rating, except to state that it is the horsepower output the motor is designed to deliver on a continuous basis when the motor is energized with its rated voltage and frequency. Motors below 1 horsepower are referred to as fractional-horsepower motors, and motors 1 horsepower or more are called integral-horsepower motors. Document whether each of the motors you surveyed is a fractional-horsepower or integral-horsepower motor.

Motor #1 1/4 Hp Motor #2 1/2 Hp Motor #3 1 HP

LOCKED-ROTOR KVA ANALYSIS

When AC induction motors are started with their full rated frequency and voltage applied, they draw locked-rotor current for a short period of time, which is an inrush current that's many times higher than the value of the full-load running current of the motor. The intensity of the locked-rotor current is calculated with the motor's design code letter. The code letter represents a multiplier that is determined by the design characteristics of the motor that will determine the maximum kVA per horsepower that the motor is capable of drawing under locked-rotor conditions. The NEMA locked-rotor design letters are given in Table 8-1. When asked to calculate the maximum locked-rotor current of each of the motors you surveyed, use the average of the two numbers of the range for each design letter to make your calculation.

HIGH ⟷ LOW

Code	kVA/HP	Code	kVA/HP
A	0–3.14	L	9.0–9.99
B	3.15–3.54	M	10.0–11.19
C	3.55–3.99	N	11.2–12.49
D	4.0–4.49	P	12.5–13.99
E	4.5–4.99	R	14.0–15.99
F	5.0–5.59	S	16.0–17.99
G	5.6–6.29	T	18.0–19.99
H	6.3–7.09	U	20.0–22.39
J	7.1–7.99	V	22.4 and up
K	8.0–8.99		

Table 8-1 Design code letter kVA-per-HP chart

The formula for calculating locked-rotor amperes is:

$$\text{LRA} = (\text{average kVA per HP multiplier from Table 8-1} * 1{,}000 * \text{HP}) \div (\text{Voltage} * 1.732)$$

When applying this formula to a single-phase motor, the 1.732 three-phase multiplier is omitted.

Using the locked-rotor design letters of the motors you surveyed, calculate the maximum locked-rotor current of each. Remember to use the average of the multiplier numbers of the range for the motor design letter you have.

Motor #1 16-18 Motor #2 9-10 Motor #3 8-9

DUTY, OR TIME RATING, ANALYSIS

This specifies the maximum length of time the motor can operate at its rated horsepower and be able to effectively dissipate the heat generated inside the motor, without exceeding the motor's temperature rise. The NEMA duty ratings for motors are intermittent, continuous duty, and special duty. If the motor is rated for intermittent duty the maximum time it may run would be expressed as a ratio of how many minutes it may run, out of a total number of minutes of the run/rest cycle—for example, 10 minutes per hour.

When replacing a motor it is important to note the duty rating of the motor being replaced, and not replace a continuous-duty motor with an intermittent or special-duty motor. Document the duty rating of the motors you surveyed.

Motor #1 Continuous Motor #2 Continuous Motor #3 Continuous

INSULATION CLASS ANALYSIS

The insulation class identifies the maximum temperature that the insulation around the stator coil wires of the motor can tolerate before breaking down. The thermal capacity of the motor insulation is classified by NEMA letter designations A, B, F, and H. Designation A means that the motor can safely operate at 105°C, B = 130°C, F = 155°C, and H = 180°C. To convert the degrees Celsius provided by the letter designation to degrees Fahrenheit, multiply the Celsius temperature times 1.8, and then add 32. Using the letter designation on each of the motors you surveyed, and the conversion method just given, calculate the temperature limitation of each motor in degrees Fahrenheit.

Motor #1 B3 Motor #2 F Motor #3 B

AMBIENT TEMPERATURE ANALYSIS

The ambient temperature rating of the motor listed on the nameplate determines the maximum ambient temperature environment the motor may be operated in without exceeding the temperature limitations of the motor's insulation. Ambient temperature is the temperature of the existing air surrounding the motor—what the air temperature of the area is without the motor running. In general, maximum ambient temperature for motors is 40°C, but some motors are rated for different ambient temperatures. If the motors you surveyed have ambient temperatures other than 40°C, calculate the ambient temperatures for each motor in degrees Fahrenheit, and document them. If no ambient temperature is listed on the nameplate, make your calculation with 40°C.

Motor #1 40°C Motor #2 40°C Motor #3 40°C

TYPE ANALYSIS

Type identifies the kind of motor it is: general purpose, capacitor start, permanent split, permanent split capacitor run, and so forth. When replacing a motor it is especially important to note the old motor type and replace it with a like type. For example, a single-phase capacitor-start motor could not be replaced with a single-phase permanent-split motor, because the permanent-split motor would not have the starting torque to start the load. Document the types of motors you surveyed.

Motor #1 SS Motor #2 P Motor #3 S9

SERVICE FACTOR ANALYSIS

Motor service factor is the percentage of overloading above the rated horsepower the motor can handle for short periods of time, such as a mechanical load surge, without stalling. The service factor is expressed as a percentage, which can be multiplied by the motor's rated horsepower to determine the maximum horsepower output the motor is capable of. A motor with a 1.15 service factor can produce 15 percent greater horsepower for a short period of time than a motor with a 1.0 service factor rating. The calculation to determine the maximum horsepower is the nameplate rated horsepower times the service factor. For example, a 10-horsepower motor with a service factor of 1.20 would be capable of producing 10 horsepower * 1.2 = 12 horsepower. Calculate the overload horsepower for each of the motors that you surveyed. If there is no service factor listed, the motor has a service factor of 1.

Motor #1 1.0 Motor #2 1.15 Motor #3 1.4

FRAME ANALYSIS

The purpose of specifying standard motor mounting dimensions with a system of frame sizes is to facilitate substituting one manufacturer's motor for another. The frame designation ensures that motors will be interchangeable to the point that they will have the same mounting requirements, and the shaft will be the same height from the bottom of the motor base.

Generally, smaller, fractional-horsepower induction motors use a two-digit frame designator system, where the number represents how many sixteenths of an inch the center of the shaft is above the bottom of the motor base, which is dimension D in Figure 8-2. For example, a 56 frame would have a shaft height of 56/16, or 3.5 inches.

Larger, integral horsepower motors use a three-digit T-frame system, where the first two digits represents how many fourths of an inch the center of the shaft is above the bottom of the motor base, which is dimension D in the drawing in Figure 8-2. The third digit represents the number of inches there are between the mounting holes in the base, which is dimension 2F in the drawing in Figure 8-2. For example, a 326T frame would have a shaft height of 32/4, or 8 inches, and the distance between the mounting holes would be 6 inches.

Figure 8-2 Motor dimensioning drawing

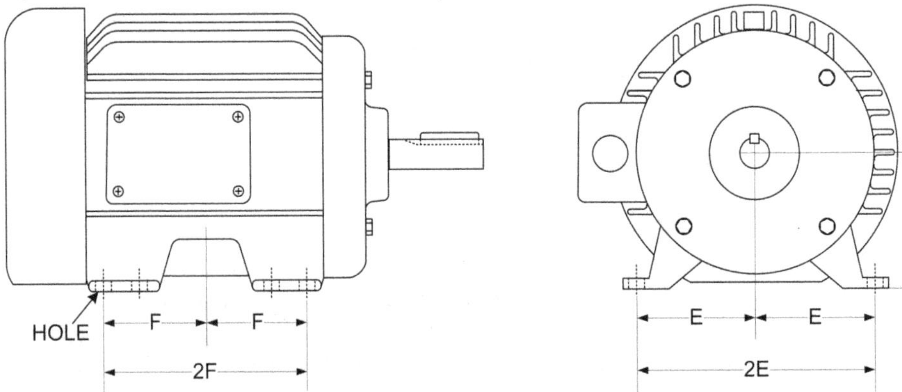

Using the frame designator systems just described, calculate the shaft height in inches of the motors you surveyed that have two-digit frame designators; calculate the shaft height and distance between the mounting holes in inches for the motors you surveyed that have three-digit frame designators. Document your calculations in Table 8-2.

Motor Frame Designation Dimension Calculations			
	Motor #1	Motor #2	Motor #3
Frame			
Shaft Height			
Distance Between Mounting Holes			

Table 8-2 Frame designator dimension calculations

After calculating the shaft height and distance between mounting holes of the motors you surveyed, measure the same dimensions on each motor to verify the frame designator. Document your measurements in Table 8-3.

Motor Frame Designation Dimension Measurements			
	Motor #1	Motor #2	Motor #3
Frame			
Shaft Height			
Distance Between Mounting Holes			

Table 8-3 Frame designator dimension measurements

DESIGN CODE ANALYSIS

Like the frame designation, NEMA has designated specific designs of motor operating characteristics for general-purpose motors, to make them interchangeable between motor manufacturers. The motor design letter describes the standardized locked-rotor torque, breakdown torque, slip, locked-rotor starting current, and other operating characteristics for each particular letter designation. The four most common NEMA standard designs for induction motors are A, B, C, and D. The majority of the motors construction electricians will encounter in the field will be design B motors, because they are considered to have the best balance between locked-rotor starting torque, locked-rotor starting current, breakdown torque, and slip. When replacing a motor it is important to note the motor's design code letter, and replace it with a like design code letter motor, because the motor application may require a specific operating characteristic of that particular motor design. Document the design code letters of the motors you surveyed.

Motor #1 B Motor #2 B Motor #3 B

ENCLOSURE ANALYSIS

This designation classifies the motor as to its degree of protection for how it protects people from accidently contacting dangerous parts, to environmental considerations such as dust and water, to methods of cooling. NEMA design motor enclosures use a variety of different descriptive designators to describe their level of protection, such as Open, Drip Proof, Guarded, Totally Enclosed Fan Cooled, Explosion Proof, Water Proof, and so forth, all of which fall into two general categories of open and totally enclosed types. When replacing a motor it is important to note the enclosure type and replace it with a motor of like enclosure type. Sometimes, however, in the case where an open-enclosure type of motor failed because dirt and debris accumulated inside the motor, it may be necessary to replace the motor with a totally enclosed type of motor. Document the enclosure types of the motors you surveyed.

Motor #1 ________ Motor #2 ________ Motor #3 ________

BEARINGS ANALYSIS

Different designations are used by different manufacturers to identify the bearing on each end of the motor. Some of the most common are DE/NDE (Drive-End/Non-Drive-End), DE/ODE (Drive-End/Opposite-Drive-End), PE/NPE (Pulley-End/Non-Pulley-End), and SE/OSE (Shaft-End/Opposite-Shaft-End); and there are many other variations on the same method.

The reason for documenting the motor bearing numbers on the nameplate is so that the motor does not have to be disassembled to get the bearing numbers if they have to be replaced. This way, the motor may remain in service while the bearings are ordered, and the motor does not have to be removed from service until the bearings are in-hand, which will reduce the motor's downtime. Document the bearing numbers of the motors you surveyed.

Motor #1 ________ Motor #2 ________ Motor #3 6285-22

MODEL AND SERIAL NUMBER ANALYSIS

The model number is important when trying to replace an existing motor, because there may be modifications or enhancements with specific models that are important to replace in the new motor. Also, the serial number or identification number may be useful when dealing with the manufacturer, and may be necessary to identify manufacturing changes that could affect replacement parts.

LAB EXERCISE #9

Disassemble a Magnetic Relay, and Contactor

Name:	Date:	Grade:

Purpose

The purpose of this lab is to familiarize the learner with the different parts and operation of electromagnetic relays and contactors, by disassembling and examining the component parts of each.

Materials:

Ice cube relay

Contactor

Clamp meter

Hand tools

MAGNETIC RELAY

Relays are sometimes referred to as *electrically operated switches*, because they are electrically controlled mechanical devices that use the power of a control circuit to produce a mechanical motion to open and close electrical contacts in one or more completely different, usually higher-power, electrical circuits. The relay control coil circuit is completely separate and electrically isolated from the relay contacts that are switched. Relays normally have relatively low current-carrying capacity ratings, which limits them mainly to control circuit applications rather than directly controlling large electrical power loads.

Ice cube relays, often just called cube relays, are relatively simple devices that are easily disassembled and reassembled without causing damage to the device if they are treated carefully. This hands-on practice in disassembling and reassembling a relay will provide practical experience and allow the learner to physically examine each individual part of the relay, and how all of the parts function together as a unit. The relay demonstrated in this lab is an inexpensive four-pole, double-throw Automation Direct relay, pictured in Figure 9-1, but just about any relay will provide the same observations.

Figure 9-1 Front of automation direct relay

The advantage of using ice cube relays is that they are normally not permanently hardwired directly into electrical control circuits that electricians encounter. Instead, the electrical connections to the internal parts of the relay are brought out to terminals on the back of the relay, as shown in Figure 9-2, which plug into a relay base, or socket, shown in Figure 9-3, which is permanently wired directly into the electrical circuit. The relay and base approach for wiring allows the relay to be removed easily from the relay base for troubleshooting or replacement, without disturbing any permanent circuit wiring.

Figure 9-2 Back of automation direct relay

Figure 9-3 Automation direct relay base

Step 1

Remove the cover from the relay. Some relays may require screws to be removed to accomplish this, but most ice cube relays simply require a small screwdriver to pry part of the cover past a catch on the relay back plate. The drawing in Figure 9-4 shows the component parts of a magnetic ice cube relay with the cover removed. The movable part of the relay is called the armature, which pivots on a hinge on one end, and is held away from the core on the other end by a spring when the control coil is de-energized. With the armature hinged on one end, the movable contacts create a single-point, make-and-break design. The armature is made of iron, with the movable contacts attached in a manner that electrically isolates them from the armature.

Figure 9-4 Automation direct relay without cover

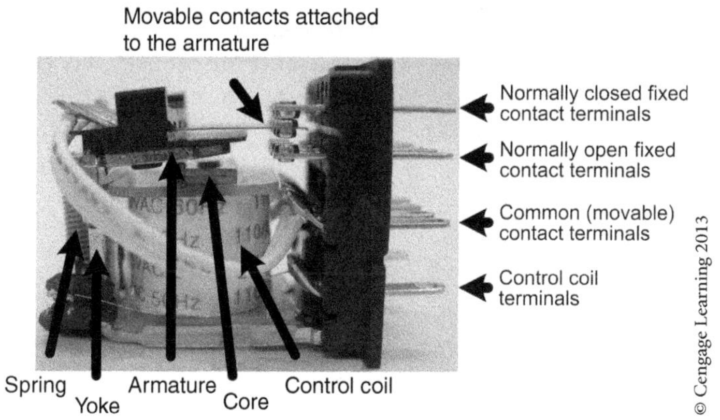

The yoke provides the physical structure that holds the relay together, and is made of iron to provide a low-permeability path to guide the magnetic lines of flux of the control coil. Figure 9-5 shows the same relay with the iron path identified with a dotted white line. The relay parts that make up the iron path for the magnetic flux to follow include the iron core through the middle of the control coil, across the iron component of the movable armature, and then down and back around to the iron core through the iron yoke. The magnetic lines of flux traveling around the iron path of the relay are what cause the armature to close against the core, which causes the movable contacts attached to the armature to change switching states.

Figure 9-5 Iron path

The control coil is a coil of wire wound around the iron core, and when the control coil is energized the two become an electromagnet. The resulting magnetic lines of flux created by the current flow in the control coil windings will follow the path of best permeability around the iron path of the device. When the air gap between the core and armature is encountered by the magnetic flux, the armature will be drawn to the core, against the tension of the spring, to create a path of continuous iron and increased permeability for the magnetic lines of flux to follow. When the control coil is de-energized, the magnetic field holding the armature against the core will dissipate, because there is no current flow in the control coil, and the spring tension will return the armature to the de-energized, or relaxed, position, with an air gap between the core and armature.

Step 2

Notice the air gap between the iron core and the armature, and notice that the movable contacts attached to the armature are making contact with one set of fixed contacts. Press down on the armature until it makes contact with the iron core, and notice how the movable contacts attached to the armature change from making contact with one set of fixed contacts to making contact with the other set of fixed contacts. This contact configuration is called the change-over, or transfer, configuration. When the armature is in the relaxed, or de-energized, position, with an air gap between the core and armature, the common movable contact on the armature makes contact with one of the fixed contacts, and breaks contact with the other fixed contact. These are called the normally closed contacts and the normally open contacts, respectively. When the coil is energized and the armature is drawn against the core, the common contact on the armature changes states between the fixed contacts, opening the normally closed contacts and closing the normally open contacts.

Once again, watch the movable contacts attached to the armature very closely as you press the armature against the iron core. Press the armature just enough to see the movable contacts make contact with the normally open set of fixed contacts. Notice that the armature has not yet made contact with the iron core; the armature has to travel further to make contact with the iron core. While watching the contacts very closely, continue pressing th armature until it makes contact with the iron core. When the armature continues to close against the iron core after the contacts have first made contact, the contacts are forced to sweep across each other. This sweeping motion is designed to clean any oxidation or debris off the face of the contacts that may have accumulated if the contacts have not been exercised for a long period of time. The additional travel of the armature also causes the arm of the movable contacts to flex slightly, and hold a steady spring tension pressure on the closed contacts.

With your clamp meter set to the continuity position, test one pole of the relay to identify the pin number of the common contact, which is the movable contact attached to the armature; the normally closed fixed contact pin number; and the normally open fixed contact pin number. For the Automation Direct relay, pin #9 is the common movable contact, pin #1 is the normally closed fixed contact, and pin #5 is the normally open fixed contact. Document your readings in Table 9-1.

Relay Pin-Out Numbers		
Common	Normally Closed	Normally Open
1,6, 11	8.4, 5	9, 3, 7

Table 9-1 Relay pole pin-out numbers

Step 3

Remove the armature spring, un-hinge the armature from the yoke, and lay the relay open, as shown in Figure 9-6. The iron core going through the center of the control coil is now clearly visible. The copper loop around the iron core is a shading ring, and is found only on relays with AC control coils. The purpose of the shading ring is to prevent the buzzing sound caused when the armature vibrates against the iron core as the AC control voltage wave form goes through its peaks and valleys, released from the iron core at the low-voltage points and drawn to the core again at the peak-voltage points. DC control coils do not need the shading ring, because DC voltage is constant.

Figure 9-6 Automation Direct Relay Laid Open

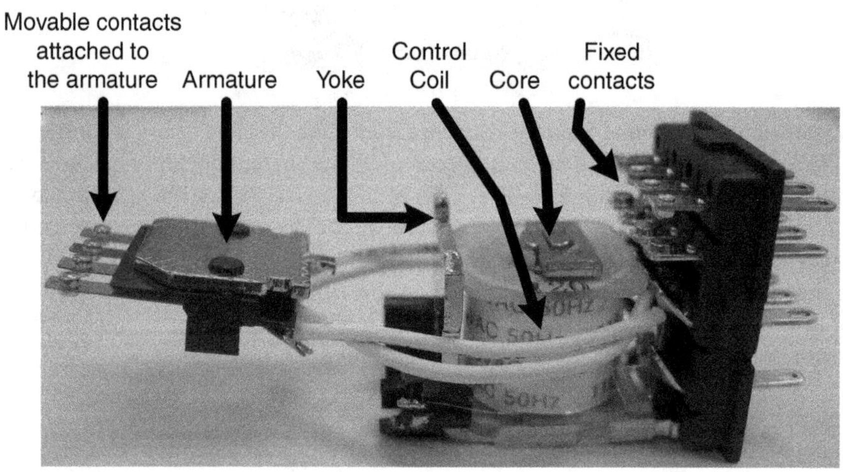

Points of observation for relays:

1. Relays normally have relatively low current-carrying-capacity ratings, which limits them mainly to control circuit applications.
2. The electrical connections to all internal parts of the relay are brought out to terminals on the back of the relay.
3. The relays themselves are normally not permanently hardwired directly into the electrical control circuits that electricians encounter.

4. The relay base, or socket, that the relay plugs into is permanently hardwired into the electrical circuit, which allows the relay to be removed easily for troubleshooting or replacement.
5. The armature is the movable part of the relay. It pivots on a hinge on one end, and is held away from the core on the other end by a spring when the control coil is de-energized.
6. Because of the hinge on one end of the armature, the movable contacts attached to the armature create a single-point, make-and-break design.
7. The movable contacts are electrically isolated from the iron armature, and each other.
8. The control coil is a coil of wire wound around the iron core, and when the control coil is energized the two become an electromagnet.
9. The control coil leads are brought out to terminals on the back plate of the relay, and the control coil is electrically isolated from all other parts of the relay.
10. The yoke provides the physical structure that holds the relay together, and with the core and armature provides the iron path of the relay.
11. The magnetic lines of flux traveling around the iron path of the relay are what cause the armature to close against the core, which causes the movable contacts attached to the armature to change switching states.
12. When relay contacts close, they create a sweeping motion that is designed to clean any oxidation or debris off the face of the contacts that may have accumulated if the contacts have not been exercised for a long period of time.
13. Relays with AC control coils have shading rings on the core to prevent the buzzing sound caused when the armature vibrates against the iron core as the AC control voltage waveform goes through its peaks and valleys.

MAGNETIC CONTACTOR

Similar to relays, contactors are also sometimes referred to as electrically operated switches, because they too use the power of a control circuit to produce a mechanical motion that opens and closes electrical contacts in a completely different, usually higher-power, electrical circuit. Unlike relays, however, contactors are designed for the power circuit, with much higher current-carrying-capacity ratings that allow them to directly control large electrical power loads.

Contactors also differ from relays in that they are designed to be disassembled and reassembled easily for alterations and repairs. Because contactors directly control high-power electrical loads, they are more likely to experience a high degree of wear and damage to certain parts. So, contactor manufacturers design the high-wear parts to be replaced easily, to extend the useful life of the contactor by repairing it rather than replacing it.

This hands-on practice in disassembling and reassembling a contactor will provide practical experience and allow the learner to physically examine each individual part of the contactor, and how all of the parts function together as a unit. A three-phase Allen-Bradley contactor, shown in Figure 9-7, is used for demonstration purposes in this lab, but just about any contactor will provide the same observations.

Figure 9-7 Allen-Bradley (A-B) general-purpose contactor

One of the first things to notice about contactors is that, unlike relays, contactors are designed to be hardwired directly into the electrical power circuit. The high currents associated with large electrical loads make using a plug-in-type base like the relay base impractical, because the terminals would have to be so large.

Step 4

Remove the shield covering the fixed and movable contacts, as shown in Figure 9-8. With the contacts exposed, push the contactor armature in and watch how the movable contacts mate with the fixed contacts. Notice the pressure spring on each of the movable contact pieces. The purpose of this spring is to compensate for any unevenness between the fixed and movable contacts, and to keep steady pressure on the contacts when they are closed, not to return the armature to the open position when the control coil is de-energized. A difference to note between the contacts of a relay and the contacts of a contactor is that relay contacts may be manufactured in the change-over, or transfer, configuration, but contactor contacts are only normally open.

Figure 9-8 A-B general-purpose contactor with contacts cover removed

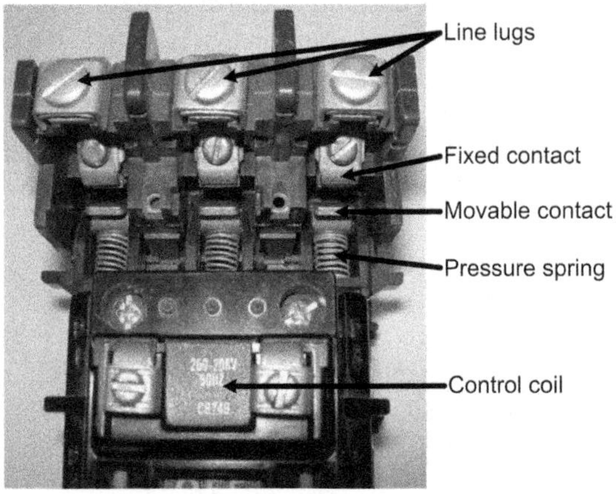

Step 5

Remove the control-coil-armature movable contacts assembly from the contactor unit, as shown in Figure 9-9, and separate it into individual pieces, as shown in Figure 9-10. Notice that the movable contacts on the armature are not hinged on one end, because contactors use a double-point, make-and-break design. The movable contacts on the armature are considered high-wear parts, so they are designed to be easily replaced. Press one of the movable contacts down against the spring tension of the pressure spring, and the movable contact will slide in and out for easy replacement.

The contactor control coil is also considered a replaceable part, and would be changed out at this point. Sometimes the control coil is damaged if it is accidently connected to the wrong control voltage, or the user may simply want to use the contactor in a control circuit with a different voltage. Contactors normally have a variety of different-voltage, interchangeable control coils to allow the contactor to be changed for a different control circuit voltage without replacing the entire contactor.

Figure 9-9 A-B control-coil-armature movable contact assembly

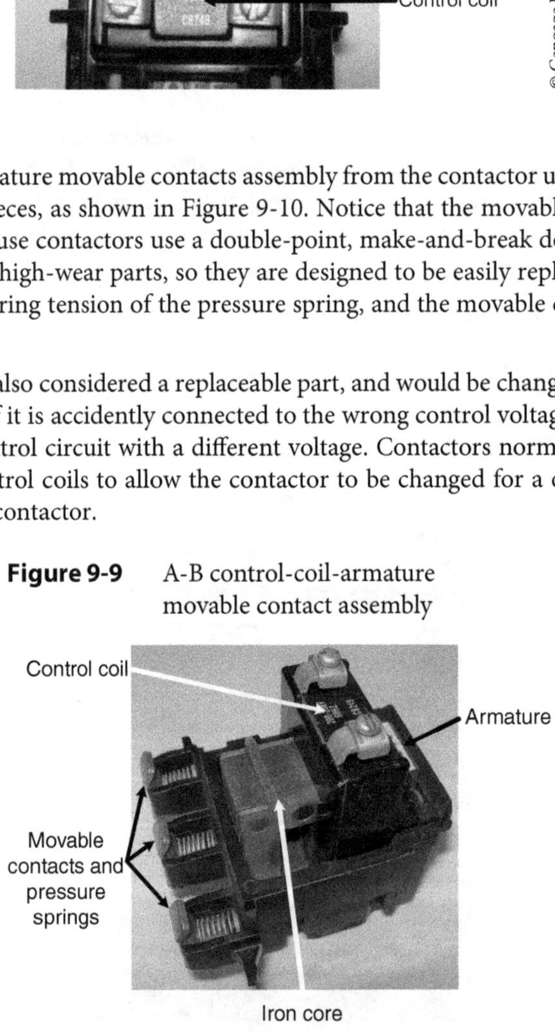

Figure 9-10 A-B control-coil-armature movable contact assembly in pieces

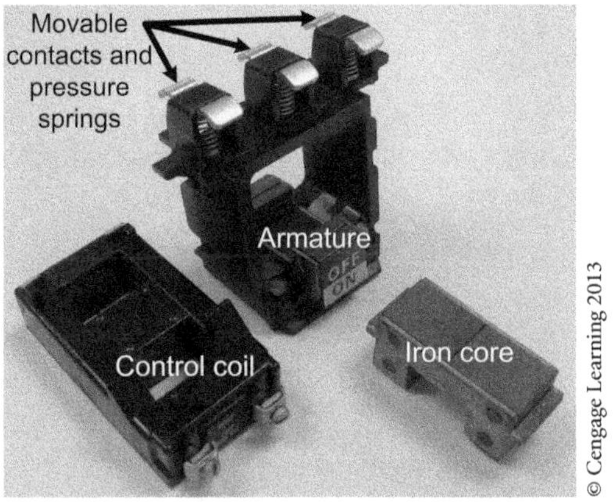

Figure 9-11 shows the fixed contact assembly, with the fixed contacts that mate with the movable contacts on the armature when the control coil is energized. Notice that the movable contacts on the armature provide the bridge between the line-side fixed contacts and the load-side fixed contacts, to energize and de-energize the electrical load connected to the contactor. The fixed contacts are also considered to be high-wear parts, so they are also designed to be easily replaced by removing the screw on each contact to remove and replace it.

Figure 9-11 A-B fixed contact assembly

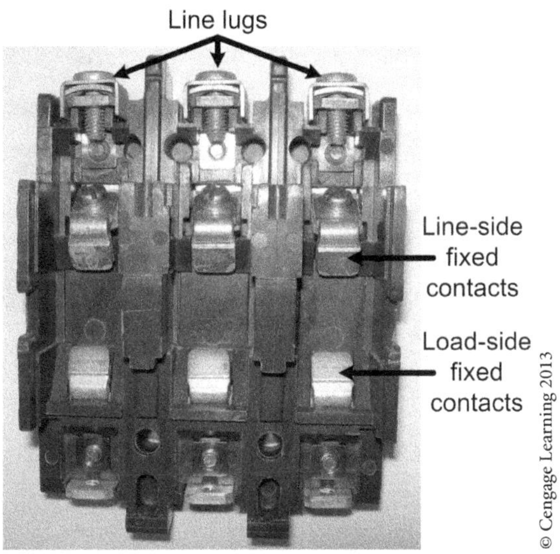

Points of observation for contactors:

1. Contactors are designed for the power circuit, with much higher current-carrying-capacity ratings that allow them to directly control large electrical power loads.
2. Contactors are designed to be disassembled and reassembled easily for alterations and repairs.
3. Contactor manufacturers design the high-wear parts to be replaced easily, to extend the useful life of the contactor by repairing it rather than replacing it.
4. Contactors are designed to be hardwired directly into the electrical power circuit.
5. Contactor movable contacts use pressure springs to compensate for any unevenness between the fixed and movable contacts, and to keep steady pressure on them when they are closed.
6. Contactor contacts are manufactured in a normally open contact design.
7. Contactors use a double-point, make-and-break design between the movable and fixed contacts.

QUESTIONS

1. Relays are sometimes referred to as what kind of switches?
2. The relay control coil circuit is completely separate and electrically isolated from what?
3. Relays normally have relatively low current-carrying-capacity ratings, which limits them mainly to what kind of electrical circuits?
4. What is the advantage of plugging a relay into a relay base, or socket, which is hardwired into the electrical circuit, rather than hardwiring the relay itself into the circuit?
5. What is the name of the movable part of a relay?
6. What mounting feature of the relay armature causes the movable contacts to have a single-point, make-and-break design?
7. What is the purpose of the iron path created by the parts of the relay?
8. What is the purpose of the armature spring in a relay?
9. What contact configuration name is given to the relay contacts that are closed when the relay is de-energized?
10. What contact configuration name is given to the relay contacts that are open when the relay is de-energized?
11. Designing the relay armature to travel further after the movable contacts initially close against the fixed contacts serves what two purposes?
12. Relays designed for AC control voltages have a shading pole on the iron core for what reason?
13. Contactors are similar to relays, except they are designed to directly control what?
14. Unlike relays, contactors are designed to be disassembled and reassembled for what reason?
15. Why are contactors designed for easy replacement of high-wear parts?
16. What is the purpose of the pressure springs on the movable contacts of contactors?
17. How many points of contact do the movable contacts of a contactor make with the fixed contacts?

LAB EXERCISE #10

Control Relays

Name:	Date:	Grade:

Purpose

The purpose of this lab is to provide hands-on practice in connecting a magnetic relay into electrical circuits and testing switching functions between terminals, to reinforce understanding of their operation. Special focus will be given to the control circuit and control coil to build a functioning stop/start station using momentary contact switches.

Materials:

11-pin octal base relay and base

2–120-volt lamps in lamp sockets (any wattage)

120-volt power supply

Hand tools

MAGNETIC RELAYS

There are basically two different types of ice cube relays that are designed to plug into relay bases, or sockets: the square base with the spade pins like the Automation Direct relay used in the previous lab, and the round octal base with the round pins like the Potter & Brumfield relay used in this lab, which is pictured in Figure 10-1. The same relay could be used for both labs if you only have access to one type of relay or the other, because the operation of both relays is the same. This lab is an opportunity, however, to introduce a second type of relay, and give you experience with a different set of pin-out terminal specifications.

Figure 10-1 Potter & Brumfield relay

The drawings for this lab assume that the relay you use will have leads from each terminal on the relay base brought out to a terminal block, similar to the drawing in Figure 10-2. For all of the drawings in this lab beyond this point, the wire leads drawn between the relay base and the terminal block will be omitted to reduce clutter in the drawing. If the relay base you are using is not wired to a terminal block, all of the numbers will still correspond with the correct terminal number for direct wiring to the relay base itself.

Figure 10-2 Potter & Brumfield relay base

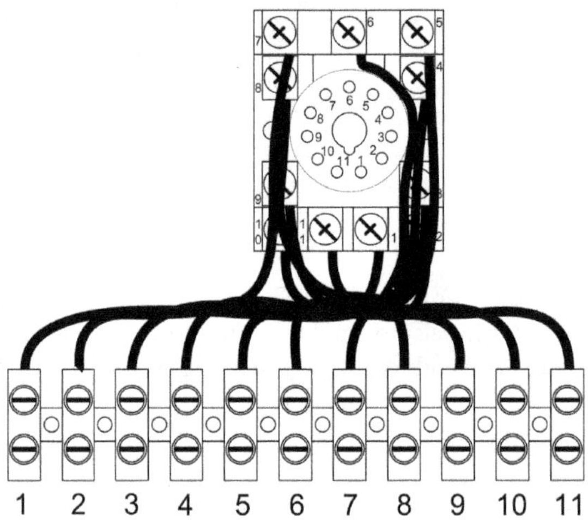

The contact pin-out configuration information for the 11-pin ice cube relay is given in Figure 10-3, for both looking at the relay base and looking at the pins on the bottom of the relay itself. Whenever working with ice cube relays it is very important to understand that when you look at the relay base the numbering is counterclockwise, and when you look at the pins on the bottom of the relay the numbering is clockwise. This can become very confusing when trying to transpose the numbers in your head, so in practice it is advisable to draw the circuit out on paper using only the base, rather than trying to switch between the two.

Figure 10-3 Potter & Brumfield pin-out drawing

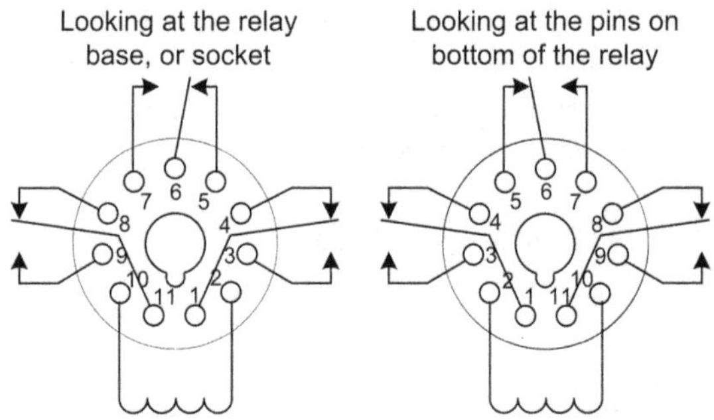

Notice that pins 5, 6, and 7 transpose between the base and the relay, and pins 5 and 7 are the only two pins that reverse function. Look at the base numbering and compare it with the relay pin numbering. Pins 1 and 11 are common points for contact circuits; pins 2 and 10 are the relay coil; pins 3 and 9 are normally open contacts; and pins 4 and 8 are normally closed contacts. The change-over, or transfer, contacts of relay terminals 1, 3, and 4 may accidently be connected to the transfer contacts of relay terminals 11, 9, and 8 and the circuit will still operate, because the terminals are mirror images of each other.

Step 1

In this step you will verify the change-over, or transfer, contacts of the relay when the control coil is energized and de-energized, using a continuity tester. Connect only the relay coil of the relay, terminal numbers 2 and 10, to 120 VAC off of Line 1 and Neutral on the 208-V terminal strip, as shown in Figure 10-4. Connect Line 1 to terminal 10 on the relay terminal strip, and connect the neutral to terminal 2 on the relay terminal strip. Do not energize the relay coil at this time. Using a clamp meter set to read continuity, complete the readings for Table 10-1.

Figure 10-4 Connect relay control coil to 120 VAC

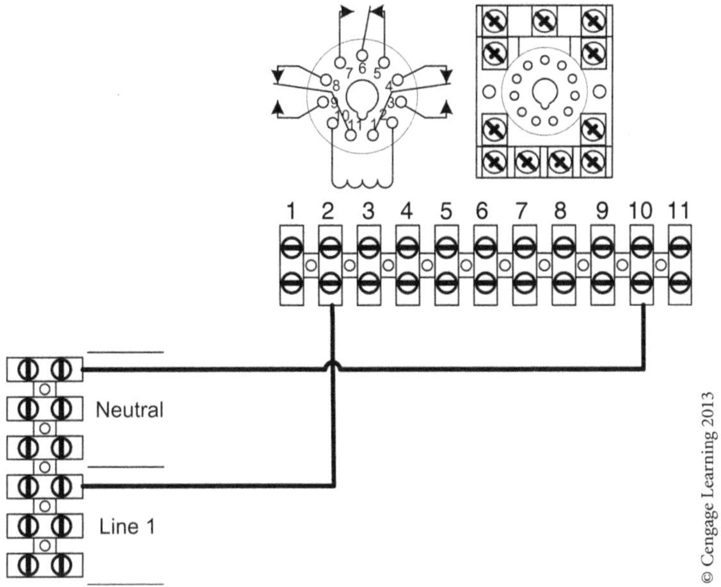

De-Energized Control Coil Readings					
Set 1		Set 2		Set 3	
Between Terminals	Reading	Between Terminals	Reading	Between Terminals	Reading
1 and 3	OL	5 and 6	0 Ω	8 and 11	0Ω
1 and 4	.1 Ω	6 and 7	OL	9 and 11	OL

Table 10-1 De-energized control coil readings

After completing the de-energized control coil readings for Table 10-1, energize the 120-VAC power supply and take the readings for the energized control coil readings for Table 10-2.

Energized Control Coil Readings					
Set 1		Set 2		Set 3	
Between Terminals	Reading	Between Terminals	Reading	Between Terminals	Reading
1 and 3	0 V	5 and 6	.1V	8 and 11	.18V
1 and 4	1.28V	6 and 7	0 V	9 and 11	0 V

Table 10-2 Energized control coil readings

1. Look at the readings for each of the three sets of change-over contacts from Tables 10-1 and 10-2; did each of the contact sets change over from continuity on one contact to continuity on the other?

   Yes

Step 2

In this step you are going to start to build a stop/start momentary contact pushbutton circuit by wiring a normally open momentary contact pushbutton in series with the relay control coil, as shown in Figure 10-5. To visually observe the change-over, or transfer, function of the relay contacts, two 120-volt lamps are also connected in the circuit—one in the normally closed contact circuit that will be lit when the relay is de-energized, and one in the normally open contact circuit that will be lit when the relay is energized. Connect the circuit and observe the change-over function when the normally open momentary contact start pushbutton is pressed and released.

Figure 10-5 Connect relay control coil with start pushbutton

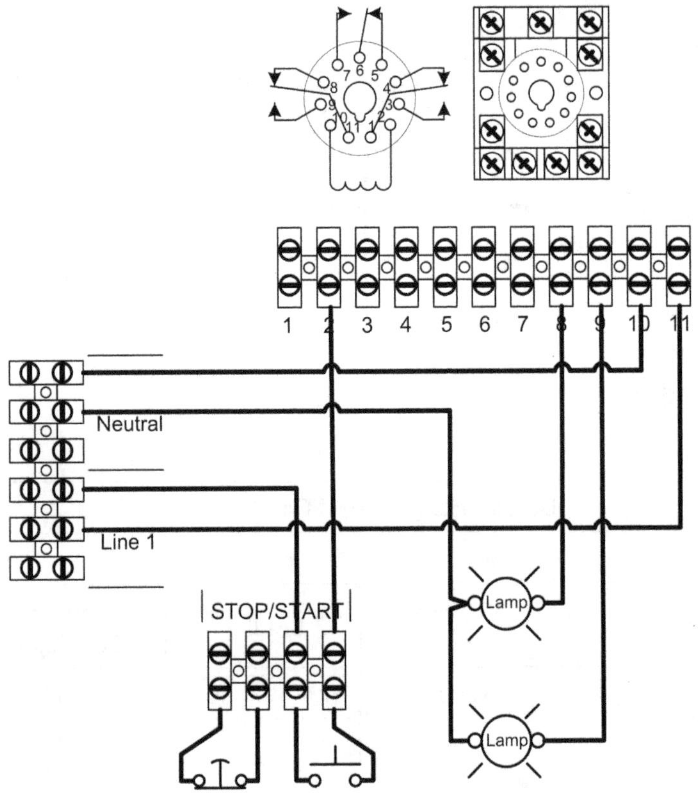

1. Do the lamps change over, or transfer, between them as the normally open momentary contact pushbutton is pressed and released? yes
2. Regardless of how you actually connected your circuit for this step, look at the circuit in Figure 10-5. Of the two conductors that are connected to Line 1, which of them is the control circuit, the top one or the bottom one? Hint: The function of the control circuit is to energize the control coil of the relay.

top is control

Step 3

In its present configuration this circuit would not be very useful in the real world. If the load were a motor, the motor would only run when the operator had his or her finger on the start button. For Step 3 you will add a holding circuit that will keep the control circuit electromagnet coil energized, which will keep the relay normally open contacts circuit, or power circuit, closed after the momentary NO pushbutton is released. To perform the holding function, another contact set is needed that is actuated by the relay armature; the contact set of pins 1, 3, and 4 will be used, as shown in Figure 10-6.

Figure 10-6 Add a holding circuit

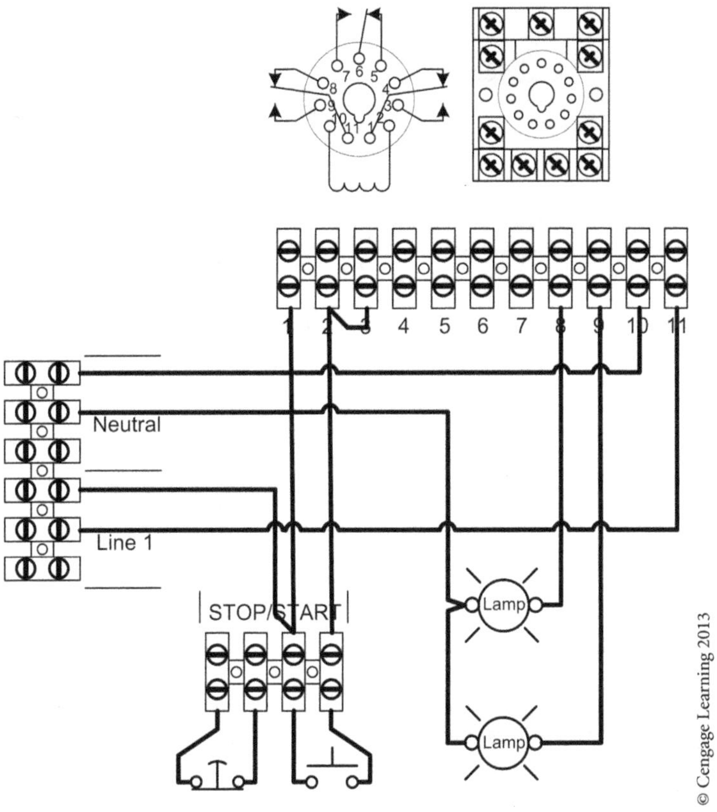

The only modification to the previous wiring should be to run a lead wire from the left terminal of the normally open momentary contact start pushbutton on the terminal strip to terminal 1 on the relay terminal strip, which is the common for the pins 1, 3, and 4 contact set on the relay; and then insert a jumper wire between terminals 2 and 3 on the relay terminal strip.

Contact group 8, 9, and 11 is the relay contact, or power circuit, and has not been modified. The control circuit, however, has been modified to bring the control power from the left terminal of the start pushbutton to the common terminal of the contact group pins 1, 3, and 4; another wire from relay terminal strip 2 to 3; and the NO contact of the contact group pins 1, 3, and 4.

Figure 10-7 shows this circuit wiring on a relay contact diagram to help understand the circuit. When the normally open momentary contact pushbutton is pressed, pin 2 of the relay becomes energized, which energizes the relay control coil. When the control coil becomes energized, the NO contacts between 1 and 3 will close, and provide a parallel path for Line 1 control power to get to coil terminal 2. Anytime after that point if the NO pushbutton is released and the pushbutton contacts open, a holding circuit through relay contacts 1 and 3 will keep the relay control coil energized, and the load powered.

Figure 10-7 Holding circuit on the relay diagram

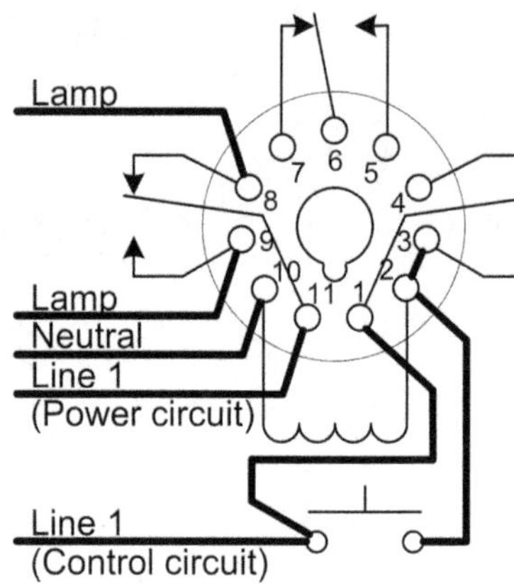

1. When the normally open momentary contact pushbutton was pressed, did the load start?
2. Did the holding circuit keep the load operating when the pushbutton was released?
3. Do you recognize a problem with this circuit? If so, what is the problem?
4. De-energize the 120-volt power source, and then re-energize it again. What happened to the circuit?

Step 4

As long as you only wanted to start a load once, and then let it run forever without stopping, the last circuit would be fine. The term *control*, however, requires that we both start and stop the load we control. At the end of the last step you learned that de-energizing the relay control coil and then reenergizing the circuit caused the circuit to reset. It is not practical to de-energize an entire process to make it stop, so another method of de-energizing the relay control coil is needed. A simple modification to the previous circuit, which is shown in Figure 10-8, will provide complete control over the load.

Figure 10-8 Complete stop-start circuit

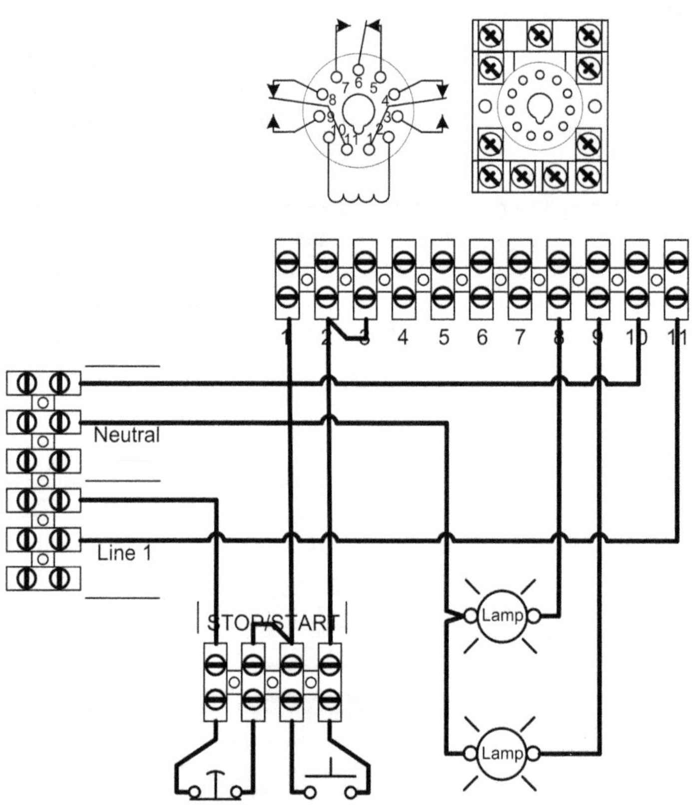

Take the lead wire presently going to Line 1 off of the 120-volt power source, and move it to the right terminal of the normally closed momentary contact stop pushbutton. Then install a new lead wire from the left terminal of the normally closed momentary contact stop pushbutton to Line 1 of the 120-volt power source. Figure 10-9 shows the wiring on a relay contact diagram to help understand the circuit.

Figure 10-9 Complete stop-start on relay diagram

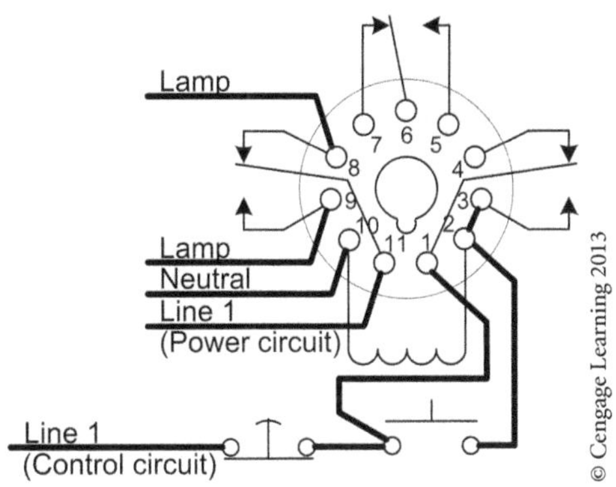

When you have completed the wiring changes, energize the 120-volt power source, and press the NO momentary contact pushbutton.

1. Did the load start? yes
2. Did the holding circuit keep the load running when the pushbutton was released? yes

Now press the NC momentary contact stop pushbutton.

1. Did the load de-energize when the stop button was pressed? yes

Step 5

A relay or contactor does not have to operate a completely different control circuit and power circuit; the control circuit and the power circuit may share the same power source, as shown in Figure 10-10. One note of caution that must be observed when connecting this circuit is that the power circuit must be tapped off before the normally closed momentary contact stop pushbutton. The electrical contacts of control circuit momentary contact pushbuttons are very small, and are intended to carry only low control circuit currents. If the power circuit is tapped off after the stop pushbutton, the circuit will still operate, but the delicate electrical contacts of the stop pushbutton will have to carry both the control circuit current and the power circuit current of the load, which would surely damage them very quickly.

Figure 10-10 Single power supply for control and power

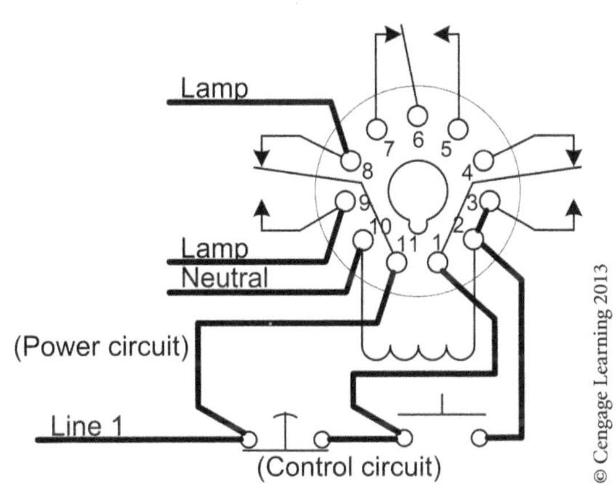

The Problem

Use the holding circuit just learned to make the following circuit. Create a circuit that turns on the lights in a tunnel when the door limit switch at the start of the tunnel is tripped by opening the door. The lights in the tunnel should remain on, even after the door at the start of the tunnel has closed behind you. When you get to the end of the tunnel, opening the door at the end of the tunnel should trip the door limit switch on that door to turn the light off. Use the diagram in Figure 10-11 to draw your wiring diagram. Use two momentary contact limit switches, one normally closed and one normally open.

Figure 10-11 Problem

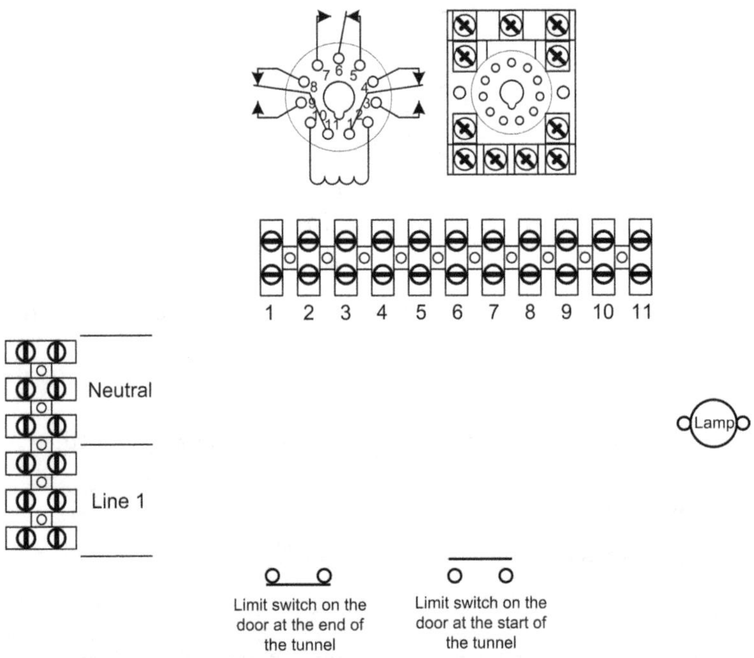

QUESTIONS

1. What two types of base configurations are ice cube relays commonly found in?
2. The 11-pin ice cube relay has how many groups, or sets, of contacts that can perform the change-over, or transfer, switching function?
3. In Step 3 of the lab, why was it necessary that the electrical contact used for the holding circuit be actuated by the relay armature?
4. What must happen to the control circuit coil to break the holding circuit contacts?
5. In Step 5 of the lab, where both the control circuit and power circuit were taken from a single line of the power supply, what precaution was given for where the power circuit is tapped?
6. What harm is caused by passing both the control circuit current and the power circuit current through the control circuit component pushbuttons?

LAB EXERCISE #11

Magnetic Motor Starter

Name:	Date:	Grade:

Purpose

The purpose of this lab is to provide hands-on practice in connecting a magnetic motor starter into electrical circuits and testing switching functions between terminals, to reinforce understanding of their operation. Special focus will be given to the electrical interconnections between the contactor and overload unit that cause them to function as a single unit.

Materials:

Motor starter

Stop/Start pushbutton control station

Three-phase motor

Multimeter

Three-phase power supply

MAGNETIC MOTOR STARTERS

Electrical circuits for powering electric motors are different from most electrical circuits. When electric motors are started with a full-voltage, across-the-line method such as a magnetic motor starter, they initially draw a very large in-rush of locked-rotor starting current, before the current drops to the motor's running current. For most motor applications, this large in-rush of locked-rotor starting current makes it difficult to protect the motor circuit conductors for both overcurrent and overload conditions at the source of their supply. In most cases the motor circuit conductors will only be protected for overcurrent at their source of supply, and overload at the motor starter. The motor starter performs two primary functions: directly start and stop the motor load, and protect the motor and circuit from overload conditions.

Magnetic motor starters are the combination of two components, a magnetic contactor unit and some type of overload protection, which work together for the express purpose of starting, stopping, and protecting electric motors. The purpose of the contactor is to directly close and open the main power circuit contacts that deliver electrical energy to the high-current motor load. The purpose of the overload unit is to sense the current draw of the motor, and open the contactor control circuit to shut down the motor and protect it against thermal overload if the motor current becomes excessive over a period of time. The two units are electrically interconnected, however, to function as a single unit in motor circuits.

The diagram in Figure 11-1 shows an Allen-Bradley motor starter with the overload unit separated from the contactor unit to make a distinction between the two. The contactor unit may be used alone to control non-motor loads, such as lighting and heating units, but when it is connected together with the overload unit on a mounting back plate, the two create a motor starter. Most manufacturers produce contactors and overload units separately, rather than building one inseparable integral unit, because there are so many non-motor starting uses for contactor units that do not require overload units.

Figure 11-1 Allen-Bradley (A-B) starter with overload unit separated

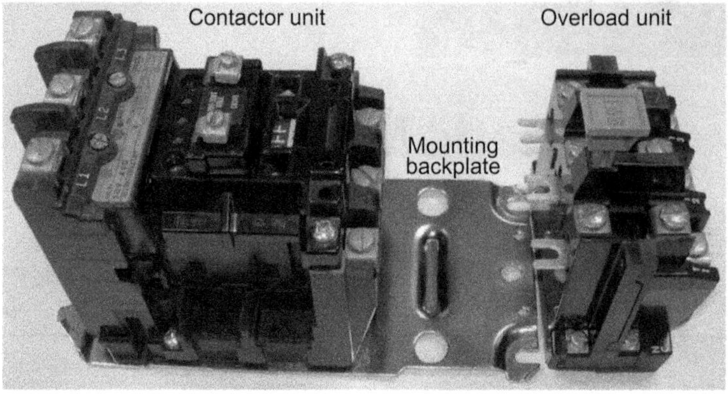

The drawing in Figure 11-2 shows the entire motor circuit, with the components of the contactor unit and the overload unit identified in separate grey boxes. The thick dark lines in the drawing represent the power circuit, which carries the high current of the motor load. The thin lines in the drawing represent the control circuit, which energizes and de-energizes the contactor control coil, but does not carry the high current of the motor load.

Notice that the power circuit goes through both the contactor and overload unit, through the main contacts of the contactor, and through the heater elements of the overload unit. The contactor unit has the main contacts that directly close and open the motor load, but the overload unit does not have the ability to directly close or open the motor load, only to sense the motor current by causing heat in the overload heaters.

The control circuit also goes through both the contactor and overload unit. On the contactor the control circuit goes through the control coil and a normally open set of auxiliary contacts with terminal numbers 2 and 3. The control coil actuates the armature, which actuates the main contactor contacts in the power circuit to energize and de-energize the motor load, and the normally open set of auxiliary contacts labeled 2 and 3 that will be used for the holding circuit. On the overload unit the control circuit goes through a normally closed set of contacts, which will open if the overload unit heaters become too hot because of higher-than-normal overload motor current.

Figure 11-2 Electrical connections between contactor and overload units

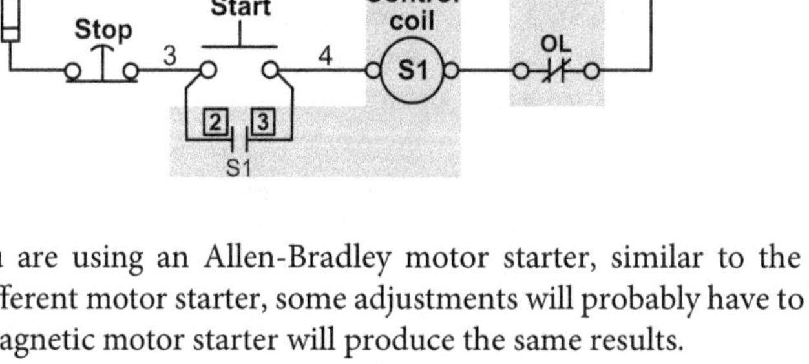

This lab assumes that you are using an Allen-Bradley motor starter, similar to the motor starter pictured in Figure 11-3. If you use a different motor starter, some adjustments will probably have to be made to the lab instruction steps, but using any magnetic motor starter will produce the same results.

Figure 11-3 A-B starter picture

Step 1

In this step you will physically locate all of the motor starter parts pertaining to the control circuit, on both the contactor unit and the overload unit. Figure 11-4 shows the control coil terminals and the holding circuit contact terminals on the contactor unit; Figure 11-5 shows the overload unit contact terminals on the left side of the unit, and the push-to-test button to manually open the overload unit contacts for testing purposes.

Figure 11-4 A-B contactor control circuit components

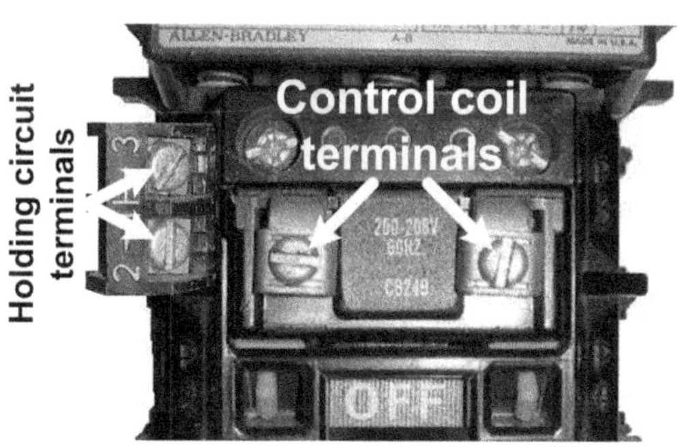

Figure 11-5 A-B overload control circuit components

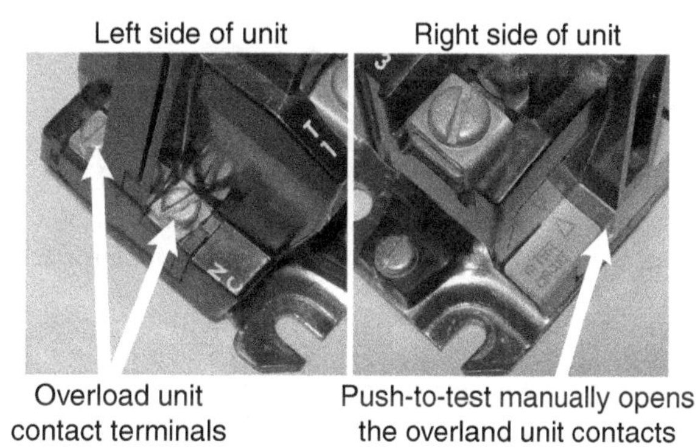

Set your multimeter to read continuity, and after testing the meter and leads for proper operation, connect the meter test leads across the normally open main contacts, from L1 to T1. With the contactor de-energized there should be no continuity between the line "L" terminals on the top of the contactor unit and the "T" terminals on the bottom of the overload unit. Manually actuate the armature by pushing it in to close the main contacts, and record your readings in Table 11-1.

Contactor Main Contacts Continuity Readings			
Non-Actuated Armature		Manually Actuated Armature	
Between Terminals	Reading	Between Terminals	Reading
L1 and T1	no	L1 and T1	yes
L2 and T2	no	L2 and T2	yes
L3 and T3	no	L3 and T3	yes

Table 11-1 Contactor main contact readings

1. Did all three of the contactor main contacts close when you manually actuated the armature?
yes

Step 2

In this step you will observe that the holding circuit terminals of the contactor unit, terminal numbers 2 and 3, are also actuated by the contactor armature. With the multimeter still set to read continuity, connect the meter test leads across the normally open holding circuit contacts. Manually actuate the armature by pushing it in to close the main contacts, and record your readings in Table 11-2.

Contactor Holding Contacts Continuity Readings			
Non-Actuated Armature		Manually Actuated Armature	
Between Terminals	Reading	Between Terminals	Reading
2 and 3	no	2 and 3	yes

Table 11-2 Contactor holding circuit contacts readings

1. Did the holding circuit contacts on the contactor unit, terminal numbers 2 and 3, close when you manually actuated the armature?

Step 3

In this step you will observe that the overload unit contacts are not actuated by the contactor armature. With the multimeter still set to read continuity, connect the meter test leads across the normally closed contacts on the overload unit. The contacts are normally closed, so the meter should immediately read continuity. If no continuity is read, the overload unit probably needs to be reset. While reading continuity through the normally closed contacts of the overload unit, manually actuate the contactor armature and record your readings in Table 11-3.

Overload Unit Contacts Continuity Readings			
Non-Actuated Armature		Manually Actuated Armature	
Between Terminals	Reading	Between Terminals	Reading
NC Contacts	yes	NC Contacts	yes

Table 11-3 Overload unit contacts readings

1. Did actuating the armature of the contactor unit have any effect on the normally closed contacts of the overload unit?

no

2. With the meter test leads connected across the normally closed contacts of the overload unit, press the push-to-test button on the right side of the unit. Does the push-to-test button open the overload unit normally closed contacts?

yes

Step 4

In this step you will verify that the overload unit can open the control circuit that energizes the contactor control coil to de-energize the motor, but cannot directly de-energize the motor itself. Connect the circuit shown in Figure 11-6, with L1, L2, and L3 connected to a three-phase power source of the correct voltage for the contactor control coil and motor, and connect T1, T2, and T3 to a three-phase motor connected for the power source voltage.

Figure 11-6 Starter wired to pushbutton station

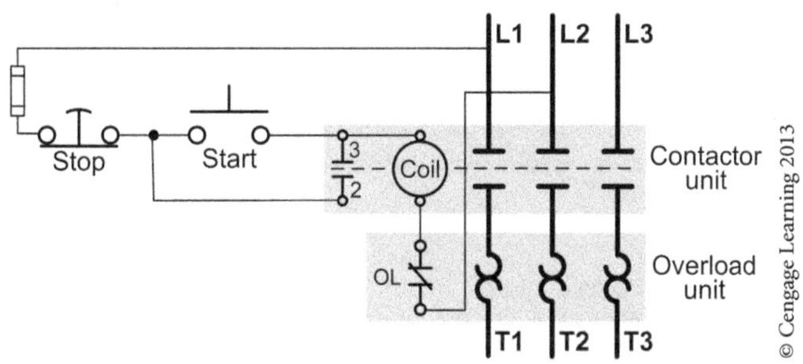

1. Start the motor by pressing the start pushbutton, and then press the push-to-test button on the overload unit. Does the motor stop?

If the motor starter you are using does not have a push-to-test button, connect a toggle switch in series with the overload unit normally closed contacts, as shown in Figure 11-7, for this step. The switch must be kept closed, and then opened to simulate the push-to-test function.

Figure 11-7 Toggle switch simulating push-to-test button

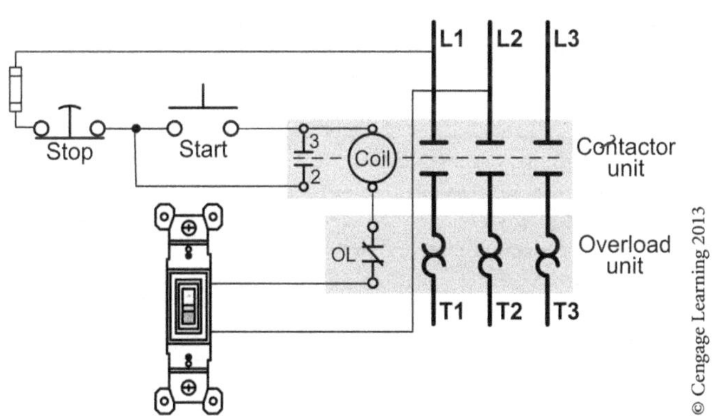

2. When you released the push-to-test button on the overload unit, did the motor automatically restart? Why?
3. Start the motor again by pressing the start pushbutton. This time, very carefully hold the contactor armature in the closed position manually, and press the push-to-test button on the overload unit. Does the motor stop this time?

The overload unit cannot directly de-energize the motor when it trips out from an overload condition; only the contactor unit can de-energize the motor. The overload unit normally closed contacts open when the overload unit trips out from an overload condition, which de-energizes the contactor control coil to de-energize the motor. That is why the motor continued to run when you held the contactor main contacts closed and simulated an overload by pressing the push-to-test button.

Step 5

In this step you will reverse the direction of rotation in which the motor you are using is turning. First, start and stop the motor, and record which direction of rotation the motor is turning.

CW / CCW

Although any two leads of the motor may be exchanged to reverse the direction of rotation of the motor, it is standard to exchange motor leads T1 and T3 on the bottom of the motor starter. Exchange motor leads T1 and T3 on the bottom of the motor starter, and restart the motor to observe the new direction of rotation.

CW / CCW

1. Did the motor reverse direction of rotation?

yes

Step 6

In this step you will measure the running current of the motor you are using for this lab. Start the motor and measure the motor current at the bottom of the motor starter on all three motor leads.

1. Was the current you read the same as the current listed on the motor nameplate? Why or why not?

no. not under load

Step 7

In this step you will choose the correct overload unit heater elements. Overload unit heater elements are sized from the motor nameplate current, or full-load amperes (FLA). Read the current from the nameplate of the motor you are using, for the voltage the motor is connected for. If you have the actual heater table for the motor starter you are using available, use it to find the correct heater element number. If you do not have the actual heater table for the motor starter you are using, use the fictitious heater table supplied in Table 11-4.

NEMA sizes 00 and 0	HEATER
0.61–0.68	H1111
0.69–0.76	H1112
0.77–0.85	H1113
0.86–0.95	H1114
0.96–1.07	H1115
1.08–1.21	H1116
1.22–1.35	H1117
1.36–1.52	H1118
1.53–1.70	H1119
1.71–1.90	H1120

Table 11-4 Overload heater chart

1. What heater element number is needed to protect the motor you used for this lab from overload conditions?

H 1115

QUESTIONS

1. What makes it difficult to protect the motor circuit conductors from both overcurrent and overload conditions at the source of their supply?
2. What conductor protection is normally provided at the source of their supply for motor circuit conductors?
3. Magnetic motor starters are the combination of what two components?

4. What is the purpose of the contactor unit in the motor starter?
5. What is the purpose of the overload unit in the motor starter?
6. Why do motor starter manufacturers produce the contactor units and overload units separately, and then combine them to create a motor starter?
7. Which circuit, the control circuit or the power circuit, is found on the contactor unit?
8. Which circuit, the control circuit or the power circuit, is found on the overload unit?
9. Which circuit, the control circuit or the power circuit, does the contactor unit have control over?
10. Which circuit, the control circuit or the power circuit, does the overload unit have control over?
11. What actuates the normally open holding circuit contacts, terminals 2 and 3, on the contactor unit?
12. What actuates the normally closed overload unit contacts?
13. How does the overload unit de-energize the motor load in the event of an overload condition?
14. Apply your critical thinking skills to the following scenario. If the power supply to the motor starter circuit is energized, but the motor is not running, manually pressing the contactor armature will cause the motor to start because the main contacts will close. Manually pressing the contactor armature will also cause the motor to lock in the holding circuit and continue to run after the armature is released, even though the start pushbutton is not pressed. What happens that causes the motor to lock in the holding circuit and continue to run?

LAB EXERCISE #12

Ladder Diagrams

Name:	Date:	Grade:

Purpose

The purpose of this lab is to provide experience in working with ladder diagrams for motor control. Special focus will be placed on identifying a few common control circuit component symbols, following ladder diagram reference documentation, and labeling ladder diagrams with reference documentation.

Material:

None needed

LADDER DIAGRAMS

Motor control circuit prints have legends that define the control symbols used in the print. Rather than focusing on memorizing a lot of different types of functional symbols, this part of the lab will focus on learning a few of the most common symbols, and spend more time on switching configurations of control circuit components. Work through the matching exercises for each of the following steps.

Step 1

In this step you will match common control circuit component symbols with their respective descriptions. In Exercise 1 (see Figure 12-1), match the number in front of the descriptions with the correct component symbol.

Figure 12-1 Control circuit components

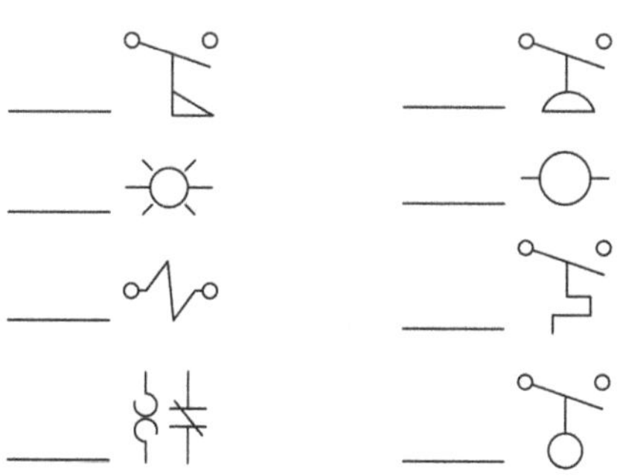

1) Pressure and vacuum
2) Solenoid
3) Temperature
4) Indicator lamp
5) Flow
6) Liquid level
7) Control coil
8) Thermal overload

Step 2

In this step you will match common control circuit contacts and switching schemes used in motor control circuits. In Exercise 2 (see Figure 12-2), match the number in front of the descriptions with the correct component symbol.

Figure 12-2 Control circuit switching components

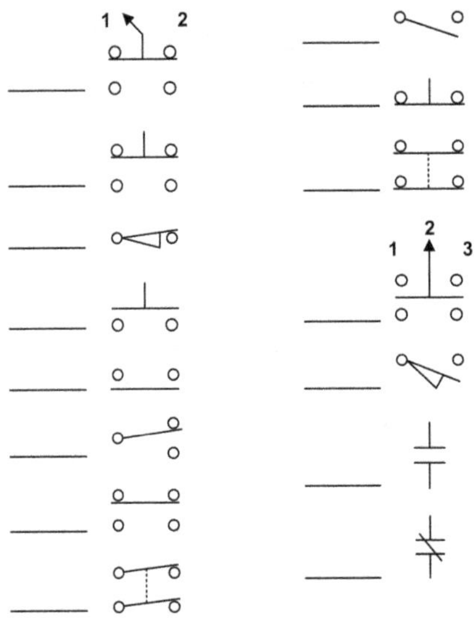

1) Normally open contact
2) DPDT double-break
3) 3-position selector switch
4) SPST single break
5) Double circuit pushbutton
6) Normally open pushbutton
7) Normally closed contact
8) SPDT double break
9) Normally open limit switch
10) DPDT single break
11) Normally closed pushbutton
12) 2-position selector switch
13) SPDT single break
14) Normally closed limit switch
15) SPST double break

Step 3

Timing relays are such an important part of motor control circuits to understand that this step is to define four different common types. For each of the timing relay symbols shown in Exercise 3 (see Figure 12-3), identify the number of the correct description for each, and identify the correct timing sequence for each. All timing sequence drawings assume a 5-second time-delay setting.

Figure 12-3 Timing relays

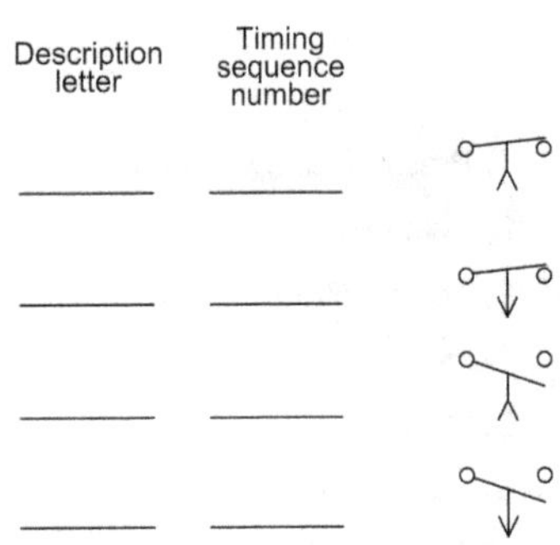

A) Timing action happens when the timing relay control coil is de-energized. The timing relay contacts close when the control coil is energized, and then time-out to reopen after a delay when the control coil is de-energized.

B) Timing action happens when the timing relay control coil is energized. The timing relay contacts are normally closed, and then time-out to open after a delay when the control coil is energized.

C) Timing action happens when the timing relay control coil is energized. The timing relay contacts are normally open, and then time-out to close after a delay when the control coil is energized.

D) Timing action happens when the timing relay control coil is de-energized. The timing relay contacts are normally closed, but open when the control coil is energized. After a delay when the control coil is de-energized, the contacts reclose.

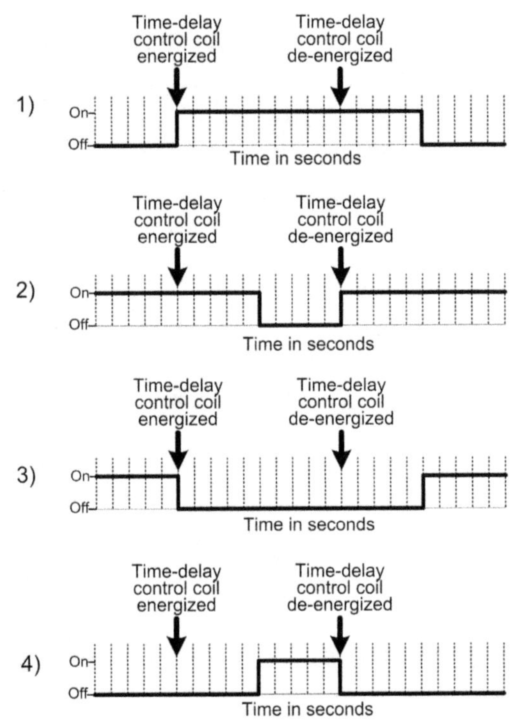

Step 4

In this step you will match the circled number reference in the ladder diagram drawing with the correct letter for the reference documentation description in Exercise 4 (see Figure 12-4).

Figure 12-4 Ladder diagram reference documentation

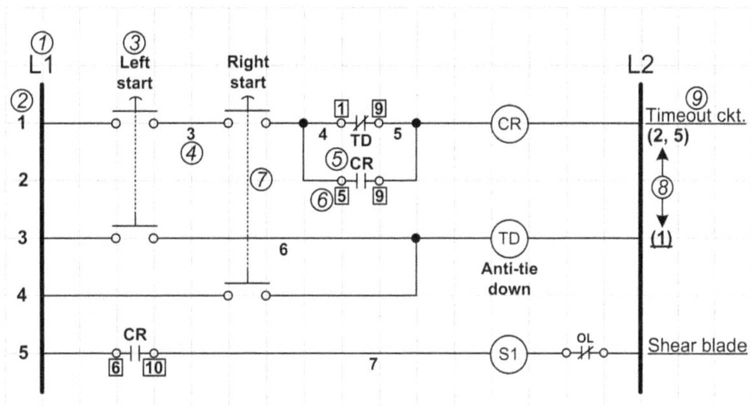

① ______
② ______
③ ______
④ ______
⑤ ______
⑥ ______
⑦ ______
⑧ ______
⑨ ______

A) Wire number
B) Contact label
C) Mechanical linkage
D) Rung number
E) Terminal number label
F) Rail label
G) Rung note
H) Component label
I) Control coil contact reference

Step 5

In this step you will complete the requested reference documentation on the ladder diagram in Exercises 5 and 6. Complete all of the reference documentation for the single rung in Exercise 5 (see Figure 12-5) for the pump 1 controller S1; then complete the missing requested documentation in Exercise 6 (see Figure 12-6).

Figure 12-5 Complete ladder diagram reference documentation—one rung

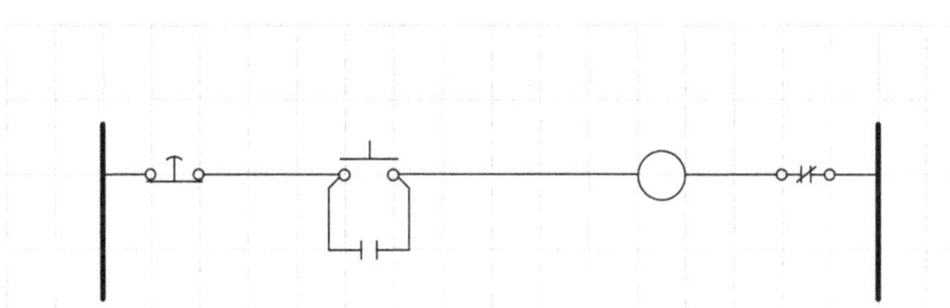

Figure 12-6 Complete ladder diagram reference documentation

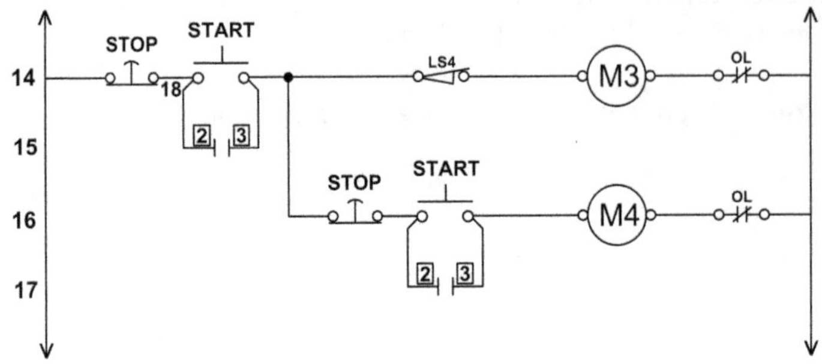

1) Continuing with the wire number 18, complete the wire numbers for this diagram.

2) The contacts numbered 2 and 3 in rung 15 are holding contacts for control coil M3, and the contacts numbered 2 and 3 in rung 17 are holding contacts for control coil M4; provide the correct reference documentation for these two contacts.

3) In addition to the contacts shown in this part of the ladder diagram, control coil M3 has a normally open contact in rung 3, and control coil M4 has a normally closed contact in rung 25; provide the correct reference documentation for these two control coils.

4) Provide a rung notes designating control M3 as conveyor #1, and M4 as conveyor #2.

Step 6

In this step you will compare the logical placement of control circuit components of a ladder diagram with the physical placement of control circuit components of a wiring diagram. Exercise 7 (see Figure 12-7) is a simple stop/start pushbutton station with a single motor starter.

Figure 12-7 Ladder diagrams and wiring diagrams

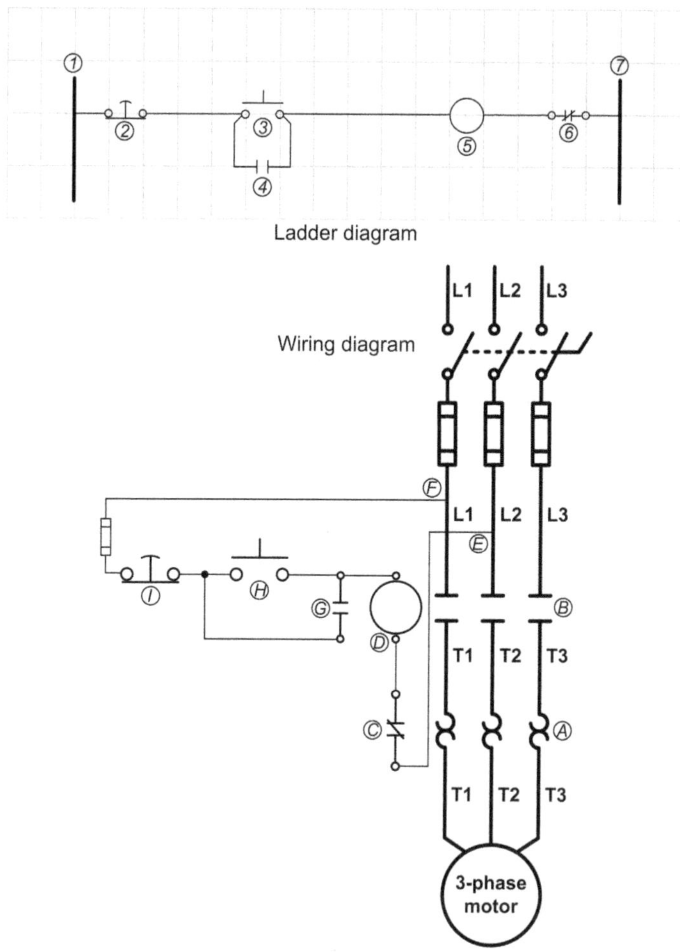

1) __________ 2) __________ 3) __________ 4) __________

5) __________ 6) __________ 7) __________

8) Where are points A and B of the wiring diagram on the ladder diagram?

LAB EXERCISE #13

Two-Wire Control

Name:	Date:	Grade:

Purpose

The purpose of this lab is to gain experience in building, observing, and understanding two-wire motor control circuits and their operation.

Materials:

Motor starter

Three-phase motor (optional)

Toggle switch, or 2-wire pilot device

TWO-WIRE MOTOR CONTROL CIRCUITS

Two-wire motor control circuits are called two-wire circuits because there are only two control circuit wires between the motor starter and the control device. Figure 13-1 is the wiring diagram drawing of a simple two-wire control circuit, showing a toggle switch as a substitute for the pilot device. Only two wires are needed, because two-wire control components are mechanical on-off types of switches, similar to the toggle switch in a room to turn the lights on and off. There is no holding circuit in two-wire control, so the holding contacts are not used.

Figure 13-1 Two-wire control circuit diagram

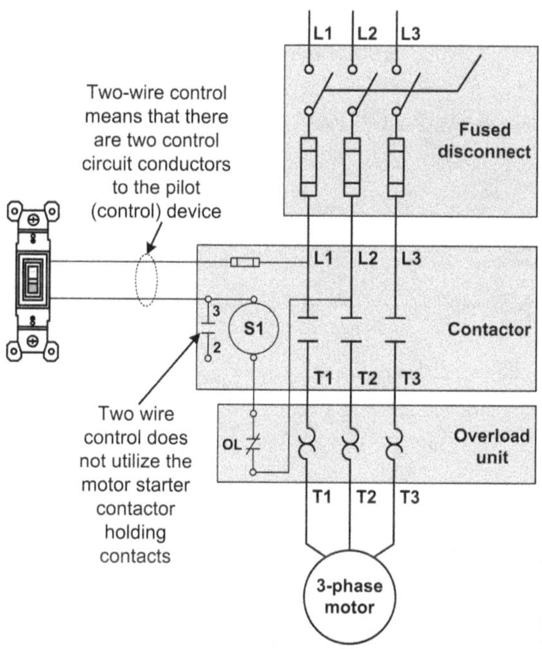

Step 1

No-voltage and low-voltage dropout protection means that if the motor starter contactor drops out for any reason, such as a power outage or severe control circuit voltage sag, the motor will drop out, and not restart again after the power is restored without operator intervention. In this step you will verify that two-wire control does not have low-voltage or no-voltage dropout protection. Figure 13-2 shows the same control circuit drawn as a ladder diagram, showing a pressure tank switch as the pilot device. Connect this circuit as shown, using a toggle switch as a substitute for the pressure tank switch. After energizing the power supply, switch the toggle switch on and off to simulate a pressure tank switch turning the motor on and off.

Figure 13-2 Two-wire control ladder diagram

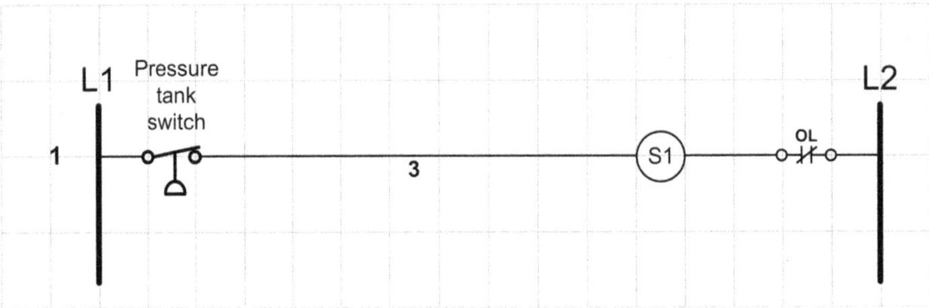

With the toggle switch turned on and the motor running, de-energize the power supply to simulate a power outage. Notice that when you reenergize the power supply the motor automatically restarts. Automatic motor restart is not a problem in most cases where two-wire motor control is used, in fact in some cases it can be an asset (such as a sump pump). If an unexpected restart of the motor would cause a safety concern, however, the motor disconnect switch should be opened in the event of a power outage and then restart the motor under controlled conditions when the power is restored.

Step 2

Probably the most common type of two-wire motor control circuit is the hand-off-automatic (HOA) control. The HOA circuit provides the same pilot device control of the circuit when the selector switch is in the automatic position, but it adds operator control for two other control options. If the HOA selector switch is turned to the hand position, the motor should start, regardless of the status of the pilot device; and if the selector switch is turned to the off position, the motor should remain off, regardless of the status of the pilot device.

In this step you will verify the operation of an HOA motor control circuit. Connect the circuit shown in Figure 13-3, substituting a toggle switch for the pressure tank switch.

Figure 13-3 Hand-off-automatic (HOA) circuit ladder diagram

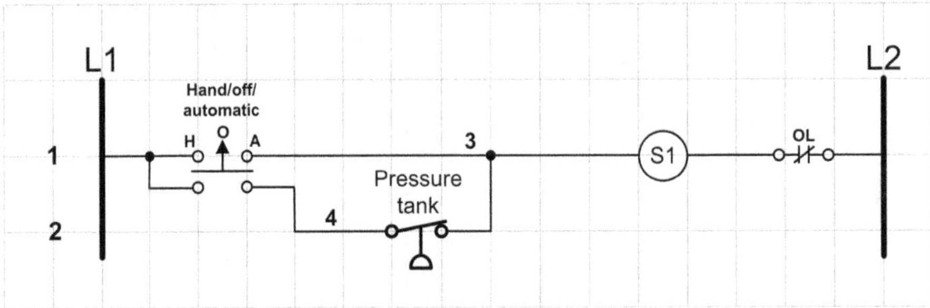

Turn the HOA selector switch to the center, or off, position, and energize the power source for the circuit. Turn the toggle switch on and off to verify that when the selector switch is in the off position the motor cannot be started, even with the pilot device.

Turn the HOA selector switch to the hand position, and notice that the motor is immediately energized. Turn the toggle switch on and off to verify that when the selector switch is in the hand position the motor runs, regardless of the pilot device status.

Turn the HOA selector switch to the automatic position, and notice that the motor is energized and de-energized by the control of the toggle switch being used to simulate a pilot device, such as a pressure tank switch.

Using the HOA ladder diagram in Figure 13-3, complete the HOA wiring diagram in Figure 13-4.

Figure 13-4 HOA wiring diagram drawing

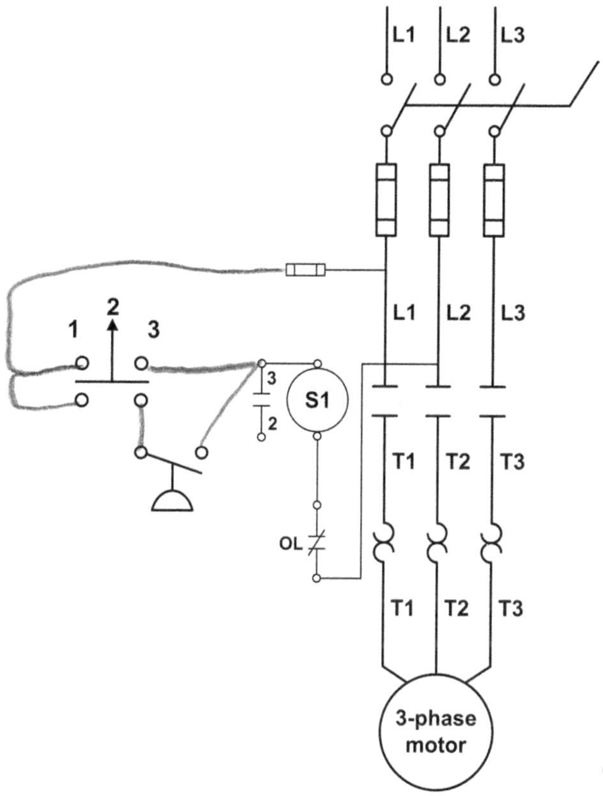

Step 3

Practice drawing the pressure tank switch pilot device and the HOA two-wire control circuits. Use the gray grids provided to keep your drawings in proportion.

Drawing Name: ______________________________

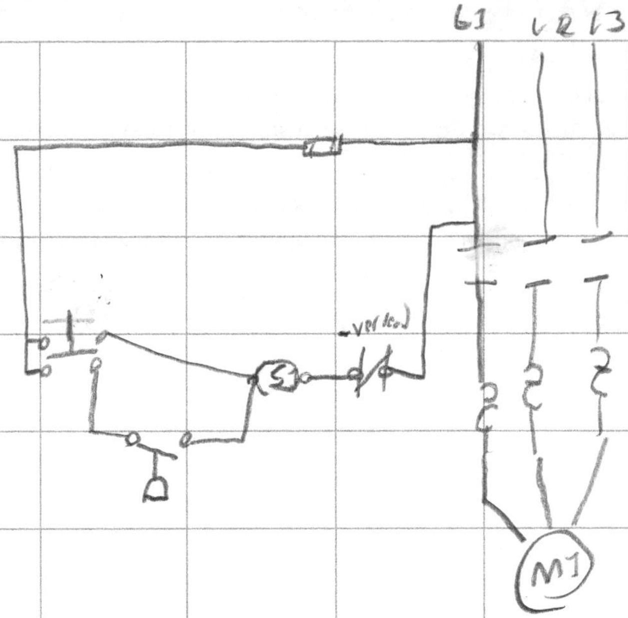

Drawing Name: ______________________________

QUESTIONS

1. What part of the contactor control circuit is not used with two-wire control? 2 and 3 terminal
2. What are the control devices called in two-wire control circuits?
3. Do two-wire motor control circuits have low-voltage and no-voltage dropout protection?
4. If it is unacceptable to have the motor restart unexpectedly when the power is restored, what step should be taken with two-wire motor control circuits in the event of a power outage?
5. In which of the control positions of an HOA circuit does the motor remain off, regardless of the status of the pilot control device?
6. In which of the control positions of an HOA circuit does the motor start and stop under the control of the pilot control device?
7. In which of the control positions of an HOA circuit does the motor start and run, regardless of the status of the pilot control device?

LAB EXERCISE #14

Three-Wire Control

Name:	Date:	Grade:

Purpose

The purpose of this lab is to gain experience in building, observing, and understanding three-wire motor control circuits and their operation.

Materials:

Motor starter

Three-phase motor (optional)

Three momentary contact stop-start pushbutton stations

THREE-WIRE MOTOR CONTROL CIRCUITS

Three-wire motor control circuits are called three-wire circuits because there are three control circuit wires between the motor starter and the momentary contact pushbutton control device. Figure 14-1 is the wiring diagram drawing of a simple three-wire control circuit, showing how the momentary contact pushbuttons of the stop/start station work in conjunction with the holding contacts of the motor starter contactor unit.

Figure 14-1 Three-wire control circuit wiring diagram

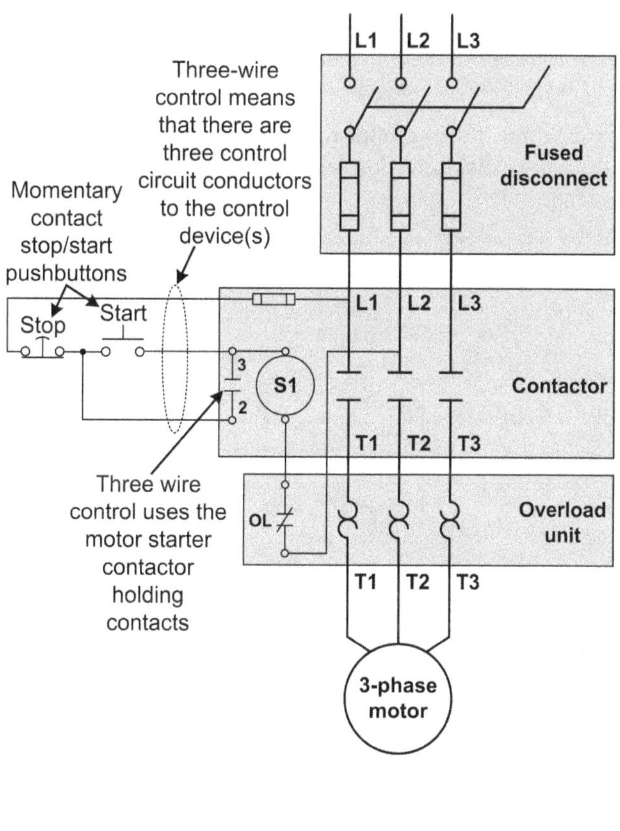

Step 1

No-voltage and low-voltage dropout protection means that if the motor starter contactor drops out for any reason, such as a power outage or severe control circuit voltage sag, the motor will drop out and not restart again without operator intervention after the power is restored. In this step you will verify that three-wire control does have low-voltage and no-voltage dropout protection. Figure 14-2 shows the same control circuit drawn as a ladder diagram. Connect this circuit as shown, using a momentary contact pushbutton station in conjunction with the motor starter contactor holding contacts.

Figure 14-2 Pushbutton control station ladder diagram

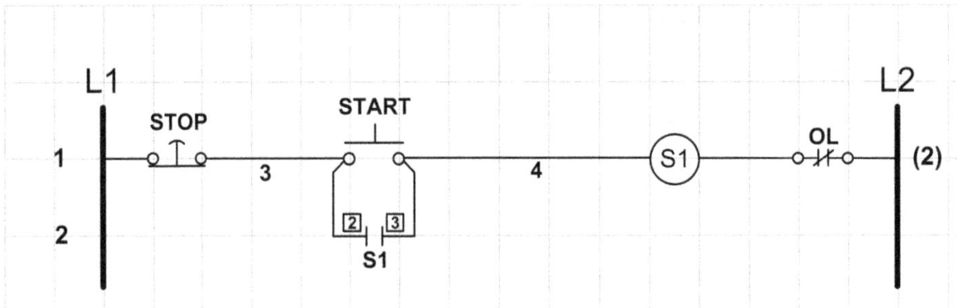

Press the normally open momentary contact start pushbutton to start the motor, and then de-energize the power supply to simulate a power outage. Notice that when you reenergize the power supply, the motor does not automatically restart. This is an example of no-voltage dropout protection. The normally open momentary contact start pushbutton opens to break the original path that energized the control coil the instant it is released; and when the power outage occurs, the motor starter contactor is de-energized, which causes the holding contacts on the contactor armature to open. Thus no more current paths remain to re-energize the motor when the power is restored.

Step 2

In this step you will control a single motor starter from two separate stop/start pushbutton control stations. Any number of stop/start pushbutton stations may be added to control the motor starter and the principles will remain the same. The control circuit power, L1, must first pass through all of the normally closed stop pushbuttons before going to a start pushbutton. After L1 has passed through all of the normally closed stop pushbuttons, it is then connected to terminal number 2 of the contactor holding contacts, and one side of all of the normally open start pushbuttons. The other sides of all of the normally open start pushbuttons are connected together, and then connected to terminal number 3 of the contactor holding contacts.

Figure 14-3 shows the ladder diagram of this circuit using two stop/start pushbutton stations, and Figure 14-4 shows the same circuit with the wiring diagram. Connect this circuit and verify that either station may be used to both start and stop the motor load.

Figure 14-3 Single motor starter with two pushbutton control stations

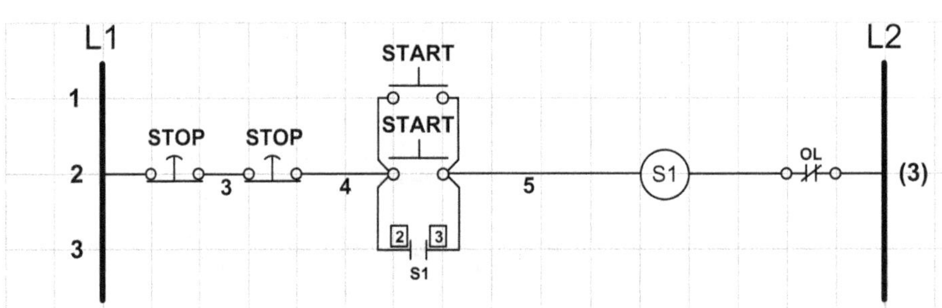

Figure 14-4 Single motor starter with two pushbutton control stations

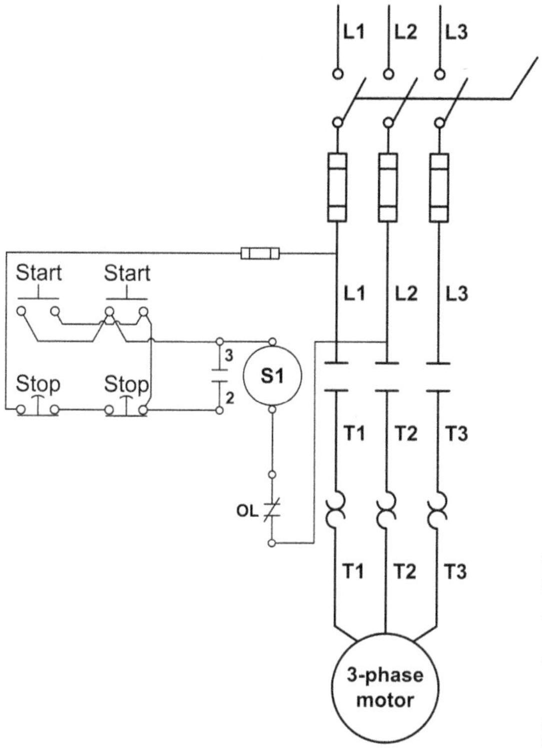

1. If the circuit is wired correctly, both normally closed stop pushbuttons should have master control over the circuit, meaning that when either is depressed the motor starter cannot be started. Hold down each of the stop pushbuttons, one at a time, and press both start pushbuttons when each stop pushbutton is depressed. Does the motor starter start when either stop pushbutton is depressed?

no

Step 3

In this step you will insert a third stop/start pushbutton station to prove you understand the principles involved in wiring multiple stop/start pushbutton stations to control a single motor starter. Once you have wired the circuit, press each of the normally open start pushbuttons one at a time, stopping the motor in between each start, to prove that each start pushbutton will energize the motor starter.

1. Hold down each of the stop pushbuttons, one at a time, and press all of the start pushbuttons one at a time when each stop pushbutton is depressed. Does the motor starter start when any of the stop pushbuttons are depressed?

no

Step 4

Chapter 14 of the text provides another 25 basic motor control circuits commonly found in the electrical industry. Connect each of them to confirm your understanding of their operation. Then practice drawing each of the circuits, using the gray grids provided to keep your drawings in proportion.

Drawing Name: NEMA Starter w/ relay

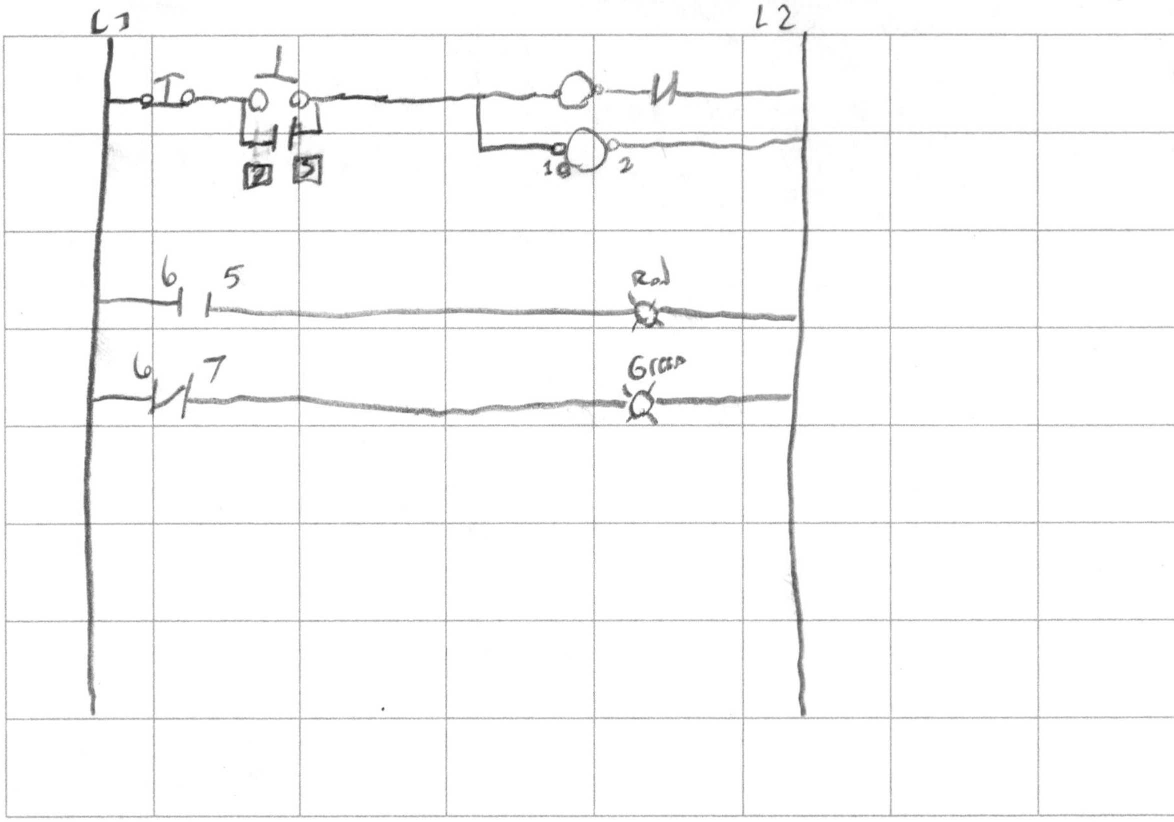

Drawing Name: IEC Starter

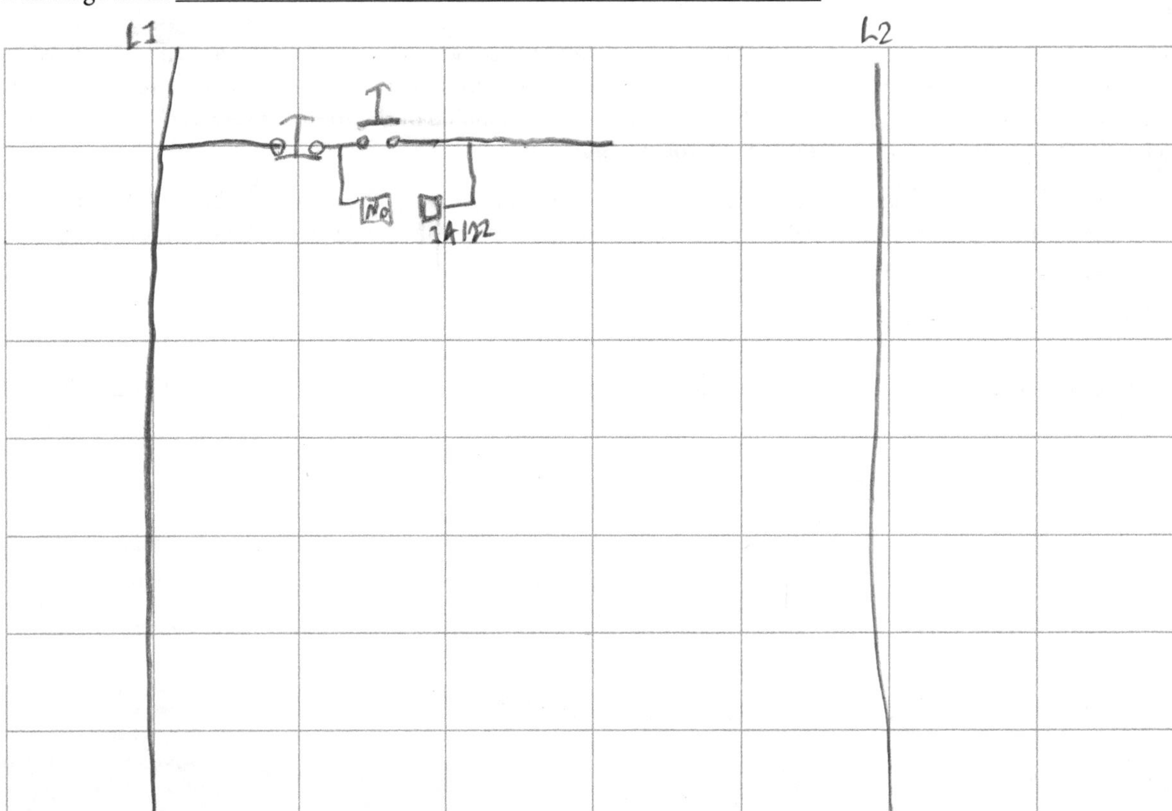

Drawing Name: ____

Drawing Name: ____

Drawing Name: ______________________________

Drawing Name: ______________________________

Drawing Name: ______________________________

Drawing Name: ______________________________

Drawing Name: ______________________________

Drawing Name: ______________________________

Drawing Name: ____

Drawing Name: ____

Drawing Name: ______________________________

Drawing Name: ______________________________

Drawing Name: ____________________

Drawing Name: ____________________

Drawing Name: ______________________________

Drawing Name: ______________________________

Drawing Name: ______________________________

Drawing Name: ______________________________

Drawing Name: ____

Drawing Name: ____

Drawing Name: ____

Drawing Name: ____

Drawing Name: ____________________

QUESTIONS

1. What part of the contactor control circuit is used in three-wire control that is not used in two-wire control?
2. How is the switching operation of the pushbuttons used in three-wire control different than the switching operation of pilot devices used in two-wire control?
3. Do three-wire motor control circuits have low-voltage and no-voltage dropout protection?

LAB EXERCISE #15

Counting Wires

Name:	Date:	Grade:

Purpose

The purpose of this lab is to understand one method of determining the number and identity of motor control circuit conductors that would be needed in the system raceways to different component physical locations.

Material:

None needed

INTRODUCTION

It is very important for the construction electrician to understand both ladder diagrams and wiring diagrams, and the different purposes they perform, to understand a particular motor control circuit. Neither of these diagrams, however, is particularly effective when trying to determine how many control circuit conductors would be needed in any particular raceway of the electrical control system. When it comes to this step, it becomes increasingly important to understand the relationship between the symbol on the electrical print and the actual physical location of each control circuit component in the real world.

When electricians are determining how to run the motor control circuit raceways of the system, they have a different mind-set than they do for any other aspect of the planning process. Certainly, a separate raceway is not going to be run to each individual pushbutton, limit switch, or solenoid from the control cabinet. The electrician will look at the physical layout of the control circuit components that must be wired, and then plan a route for the raceway that will carry the control circuit conductors for more than one control component together along the way. It is important to understand an effective, accurate method of determining the conductor size and count for each raceway run, because that information will be used to calculate raceway and junction box size, among other considerations.

Step 1

To understand this concept, the drawing in Figure 15-1 is designed for you to draw out each individual control circuit conductor, through the raceway and to each component. The ladder diagram of how the motor control circuit is actually wired is provided for your reference. Draw the individual control circuit conductors, using the raceways provided, to determine how many conductors will be needed in the raceway between the motor starter and the stop/start station, and what the wire numbers are. After drawing the control circuit conductors, draw the power circuit conductors also.

1. How many control circuit conductors are needed in the raceway between the motor starter enclosure and the stop/start pushbutton station, and what are their wire labels?

As the control circuit becomes more sophisticated, a different, more practical method must be used. One method of determining what wire number conductors must be in each raceway is to draw a rough sketch of the raceway with the control circuit components inserted along the raceway, in the order they will be wired in the field. The drawing does not have to be to scale, show actual control circuit component symbols, or even show individual conductors. A simple line for the raceway, with boxes labeled for the control circuit components along the line, is sufficient for these purposes.

Figure 15-1 Motor starter and stop/start station

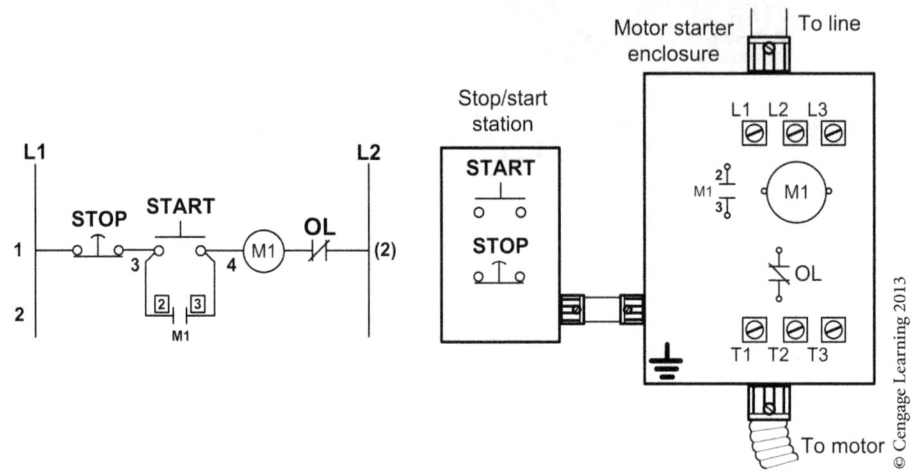

Step 2

The drawing of a single motor starter with three stop/start pushbutton stations is shown in Figure 15-2. First look at the ladder diagram of the circuit, and write the wire label numbers of the conductors that connect to the pushbuttons in each of the stop/start pushbutton stations. Next, work backward from the end of the line, back toward the motor starter, and write down each conductor wire label number that must be in each raceway for the control components along the way.

Figure 15-2 Three stop/start pushbutton stations

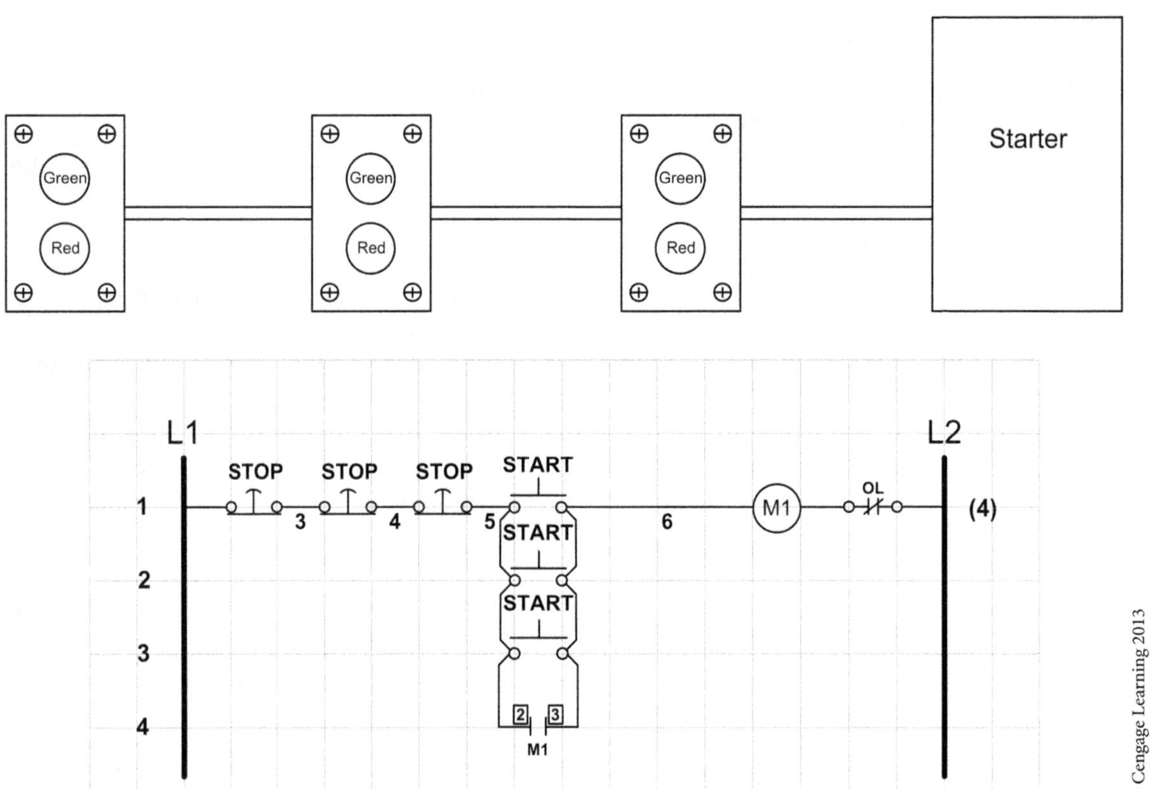

Step 3

To continue this practice on something slightly more complicated, consider the controls for a simple conveyor system shown in Figure 15-3. As shown, the conveyor has a control cabinet with the motor starter, a stop/start station, an emergency stop station, and a limit switch to stop the conveyor to prevent the product from falling on the floor. The ladder diagram shows the logical operation of the conveyor, and the drawing shows how the electrician plans to run the raceways; the wiring diagram is not shown. Practice the same process as the last step, by first writing the wire label numbers of the conductors that connect at each of the control components shown, and then working your way backward to the control cabinet.

Figure 15-3 Conveyor system

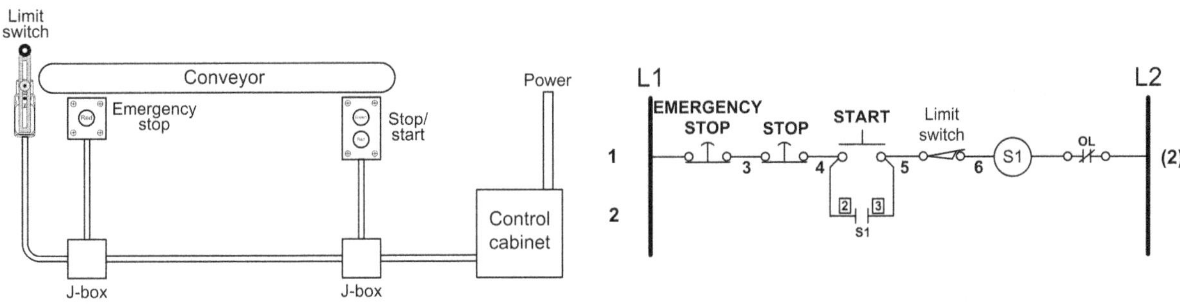

Step 4

Figure 15-4 shows a lift station example to determine what control circuit conductors have to be installed in which raceways, with the ladder diagram for the example in shown Figure 15-5. Lift stations are used to make water, usually sewage, flow uphill. The dot-dash line in the diagram is to model the grade of a hill, but does not accurately represent the actual percent grade or distance between lift stations. In reality, depending on the incline grade, the lift stations may be hundreds or thousands of feet apart. The lift station is designed to pump the water from the bottom of the sump to the outlet at the top of the sump. The pipe between the sumps is sloped downhill to the next sump uphill.

Look at the ladder diagram for this application in Figure 15-5. Lift stations #1 and #2 work exactly the same, with the exception that the CR2 NC (inhibit) contacts in rung #2 prevent lift station #1 from pumping if for some reason lift station #2 cannot keep up, and the water has risen to the alarm level. The working beacon is a green lamp visible from a distance, and the alarm beacon is a red lamp visible from a distance, so that it is easy to determine the operating condition of each sump. Assume that the control relays will be mounted in the wiring trough at each sump location when determining wire counts.

Figure 15-4 Lift station drawing

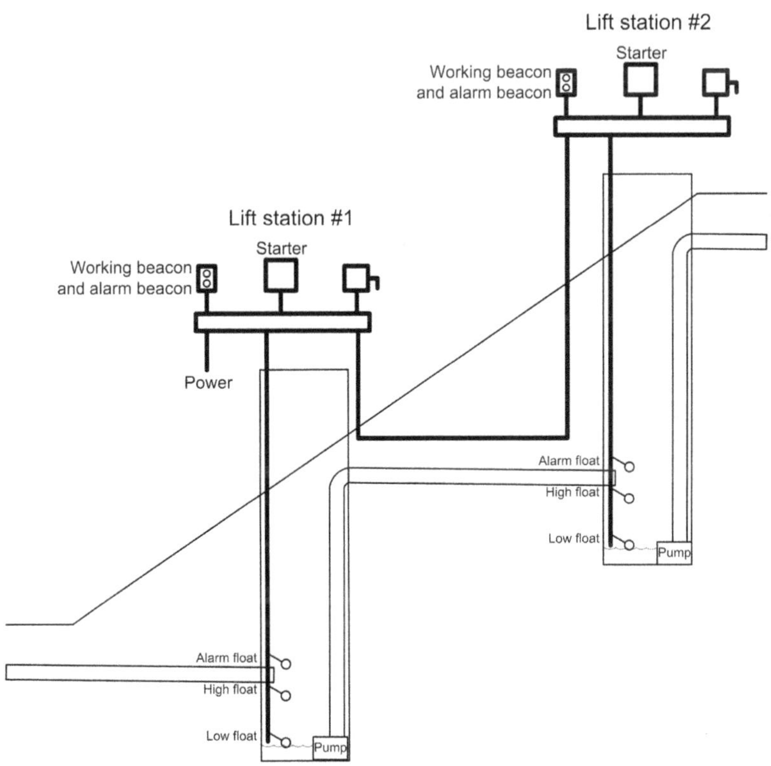

Figure 15-5 Lift station ladder diagram

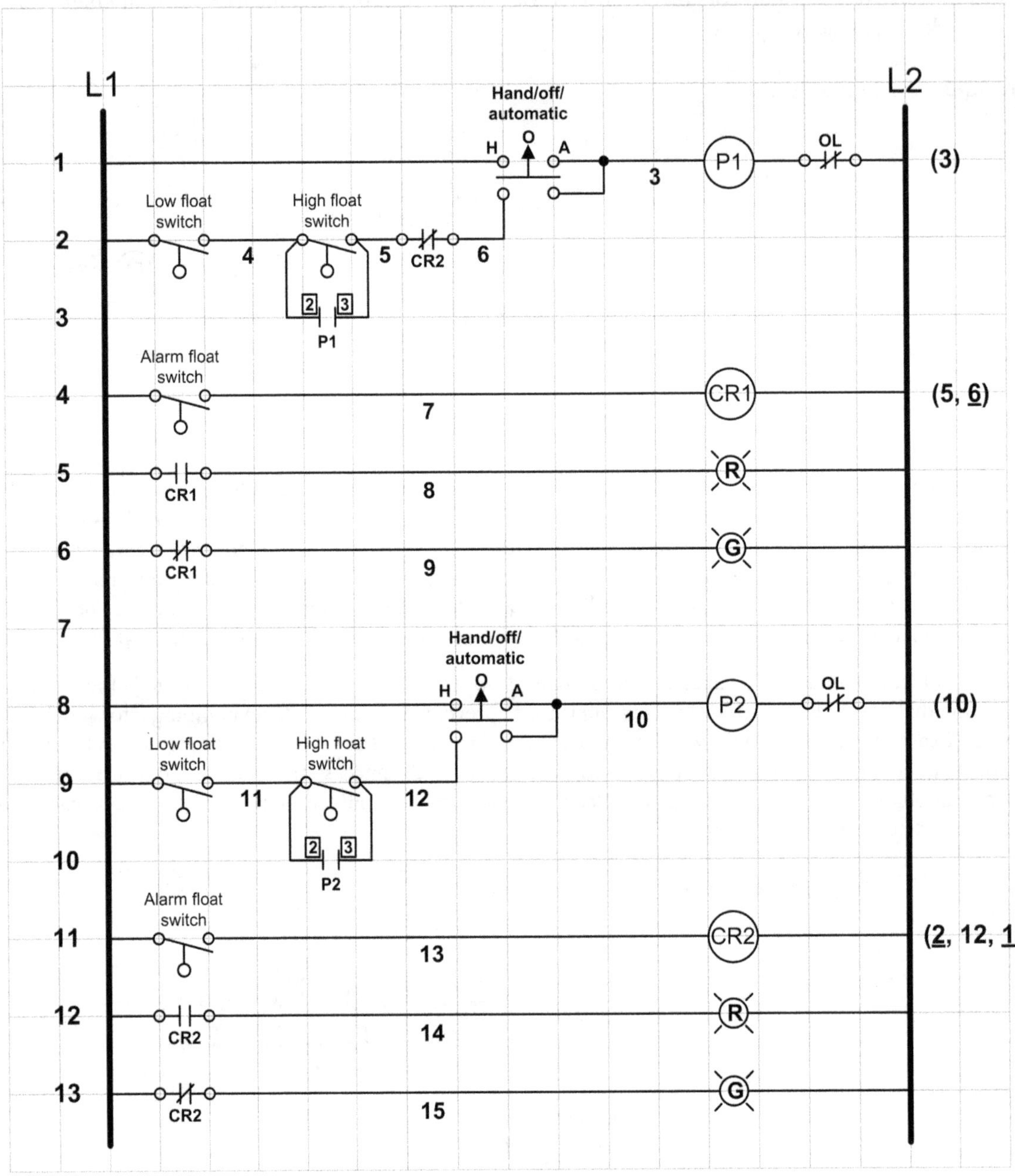

Step 5

Figure 15-6 shows a pressure washer process with the raceways laid out between each of the control components, and Figure 15-7 shows the ladder diagram for the process. In this step, determine the control circuit conductors and the wire label numbers for each of the raceways.

Figure 15-6 Pressure washer

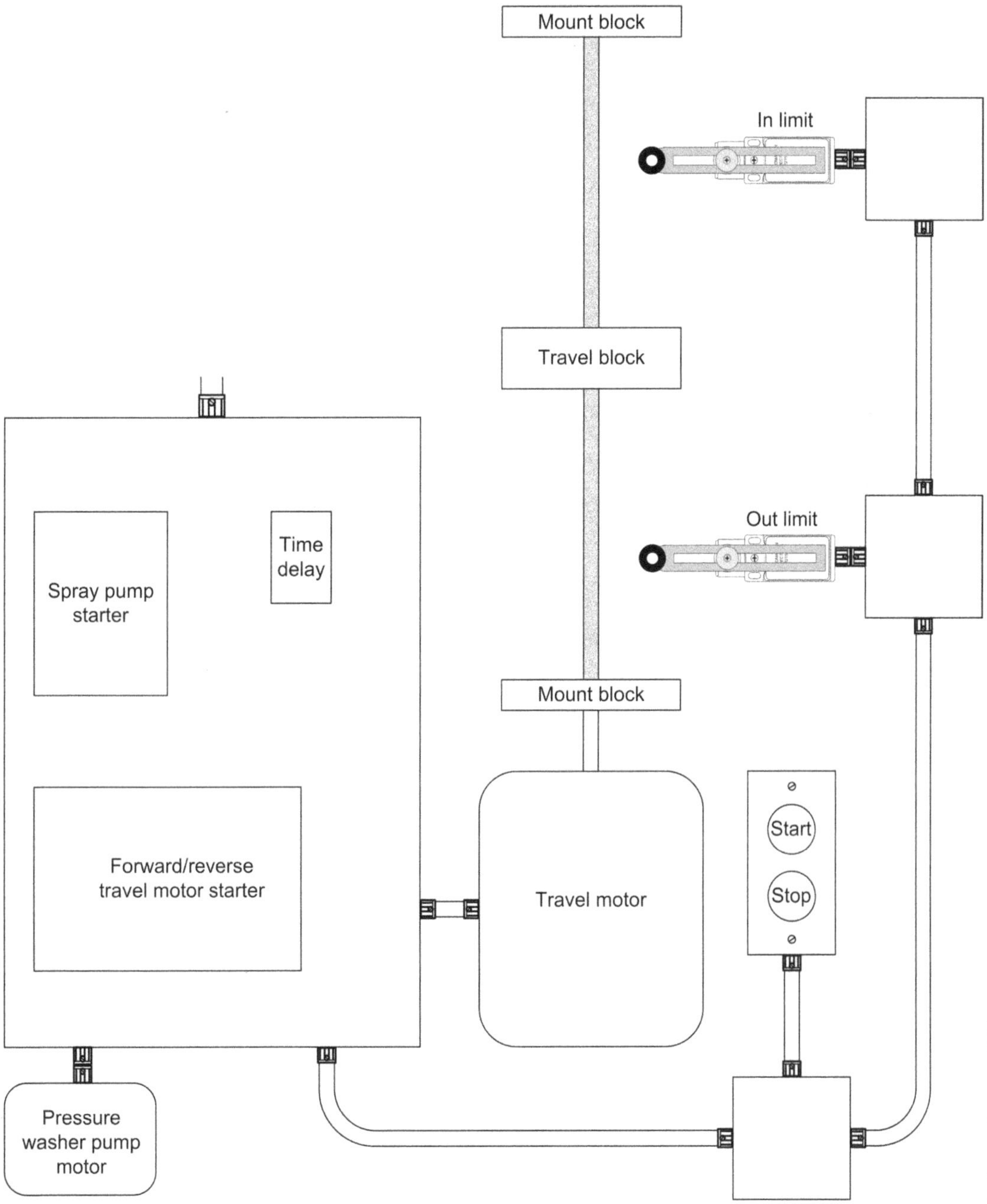

Figure 15-7 Pressure washer ladder diagram

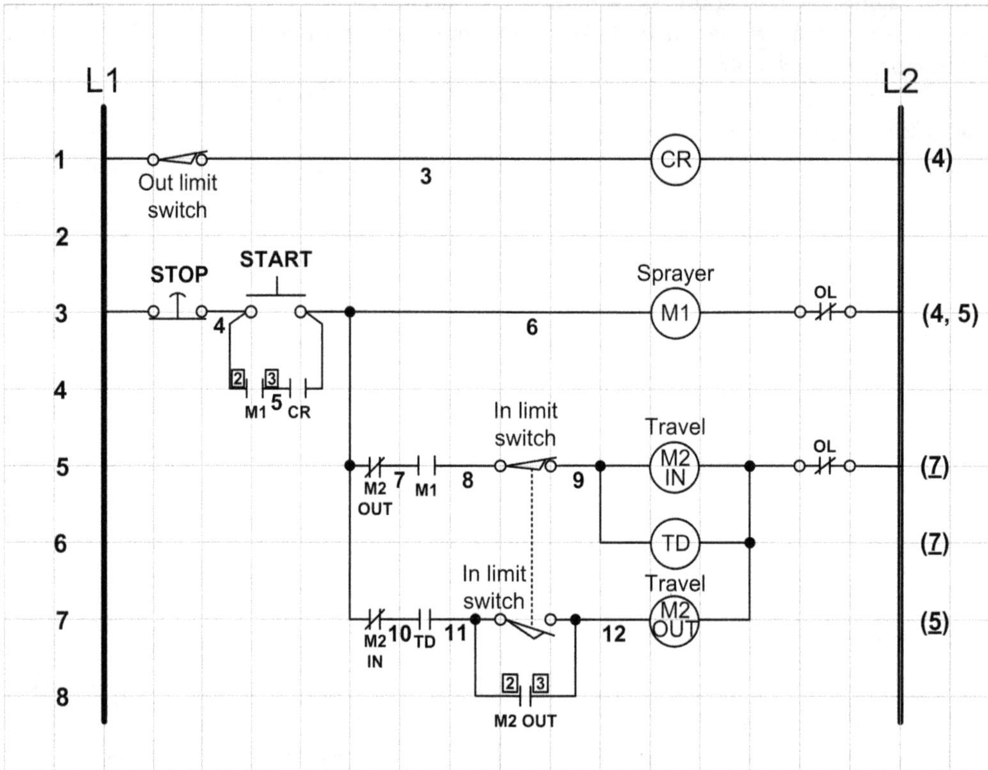

LAB EXERCISE #16

Commission an Automation Direct GS2 VFD

Name:	Date:	Grade:

Purpose

The purpose of this lab is to demonstrate the limited functionality of the Automation Direct GS2 drive with default operating values, and learn to configure the drive for basic commissioning parameters commonly encountered.

Materials:

Automation Direct GS2 VFD

Three-phase induction motor

Hand tools

Caution: Making incorrect wiring connections to the GS2 drive, such as accidentally connecting 120/208 volts to the control circuit terminals of the VFD, will permanently damage the unit. Use caution to prevent damaging the VFDs. Read and understand the wiring connections of each step fully before attempting them. For each instruction that requires action, or something to verify, a check box will appear on the right-hand side for you to check off each step as you complete it.

POWER AND MOTOR CONNECTIONS

Connect the L1, L2, and L3 terminals of the drive to the L1, L2, and L3 terminals of the 120/208 volt power source.

Check ☐

Connect the T1, T2, and T3 terminals of the drive to the T1, T2, and T3 terminals of the MOTOR #1 terminals.

Check ☐

Verify that the safety equipment ground is connected.

Check ☐

LOCAL KEYPAD COMMAND OPERATION

The GS2 drive will boot with many default operating parameters, but the most basic operating parameters of concern in this lab are listed below. Remember, default operating parameters are designed to meet the operational needs for a wide range of applications with a minimum of programming at the time of installation. These default parameters, however, may not always meet all of the operational needs of every application, and may need to be changed for some specific applications.

The following default parameters normally do not need to be changed, or are parameters that are determined by drive class, which cannot be changed:

- Display shows the output frequency of the drive
- 240 volts motor nameplate voltage
- 60 Hz motor base frequency

- 1,750 RPM motor base RPM
- 1,750 RPM motor maximum RPM

The following default parameters will likely require changes for basic commissioning, or to meet specific operating characteristics necessary for a particular application:

- 2.5 amperes motor nameplate current
- 10-second acceleration time
- 30-second deceleration time
- Constant torque operation
- Operation determined by digital keypad

1. Operation of the unprogrammed GS2
You should have the power leads (L1, L2, and L3) and the motor leads (T1, T2, and T3) connected to the VFD at this time; no external remote control wiring should be connected to the GS2. When power is initially applied, the GS2 will boot up with a predetermined set of operating parameters, as discussed earlier. By default, the drive will automatically boot to keypad operation, with the display indicating the output frequency of the drive. Apply power to the unit and notice that the drive boots and displays H 0.0 (H = output frequency).

Check ☐

a. By default the potentiometer on the front of the drive will control the output frequency, and therefore the speed, of the motor. Turn the potentiometer clockwise (CW) and counterclockwise (CCW) and notice that the motor does not turn.

Check ☐

b. Return the potentiometer to the fully CCW position. Look at the display and notice the status of the indicator light-emitting diodes (LEDs). By default, the drive booted to FWD (forward) rotation and STOP modes; these LEDs should be lit solid to indicate they are the active modes.

Check ☐

c. To release the STOP mode, push the RUN button. Two operations will happen; the green RUN LED will light, and the cooling fan will start (anticipating heat from operating a motor load).

Check ☐

Note: When the RUN button is pushed, the red STOP LED will start to blink to indicate it is not the active mode, and the green RUN LED will turn on solid to indicate it is the active mode.

d. With the drive in RUN mode, turn the potentiometer in the CW direction and note the rotation direction of the motor. Press the STOP button, and return the potentiometer to the fully CCW position and document your observation of the direction of rotation.

CW/CCW (circle one)

Check ☐

a. After the motor has stopped, push the FWD/REV button. Note that the FWD (forward) LED blinks and the REV (reverse) LED lights solid; the FWD LED will not stop blinking until the RUN button is pressed.

Check ☐

b. The drive should have a solid STOP LED to indicate that stop is the active mode. This indicates that the motor will not run until the RUN button is pushed. Push the RUN button and note the change to the drive. The RUN LED and REV LED should be on solid, and the STOP LED should be blinking.

Check ☐

c. Turn the potentiometer CW and note the rotation direction of the motor. Press the STOP button, turn the potentiometer to the fully CCW position, and return the GS2 drive to forward rotation (push FWD/REV) after your observation.

CW/CCW (circle one)

Check ☐

d. An additional feature of a VFD is that it will not allow the motor to reverse direction of rotation with the motor at operating speed; it will automatically stop the motor before reversing rotation direction. To prove this, press the RUN button, turn the potentiometer to the fully CW position, and allow the motor to attain full speed.

Check ☐

e. Without first pressing the STOP button, press the FWD/REV button and note that the VFD will stop the motor in the direction it is turning, before reversing the motor's direction of rotation and restarting the motor.

Check ☐

f. Press the STOP button when this step is completed, and return the potentiometer to the fully CCW position.

Check ☐

g. To verify that the GS2 drive default acceleration time is 10 seconds, turn the potentiometer to the fully CW position with the drive in STOP mode (STOP LED solid).

Check ☐

h. Press the RUN button, and time how long it takes the motor to reach full RPM.

Check ☐

i. Remember that the default display is output frequency, not RPM. Press the DISPLAY button once to put the display into the RPM readout.

________________ Acceleration time (seconds)

___________________ Full-speed RPM.

Check ☐

Note: After 90 seconds, the display will default back to the output frequency. To display RPM, simply press the DISPLAY button again.

j. After achieving 1,750 RPM and documenting the result in the last step, press the STOP button and time how long it takes for the motor to reach 0 RPM. Looking at the default parameters listed earlier, the deceleration time should be 30 seconds.

Check ☐

k. Return the potentiometer to the fully CCW position.

________________ Deceleration time (seconds)

Check ☐

2. GS2 operating parameter access
The GS2 operating parameters are organized into 10 different parameter groups, P0.00 through P9.00, according to their functions. Pressing the PROGRAM key on the drive's integral keypad will rotate through the individual parameter groups. Each of the 10 parameter groups contains parameter selections, which may be accessed by pressing the UP/DOWN arrow keys when the display is on the desired parameter group. Once the desired parameter selection is displayed, pressing the ENTER key will display the value stored for that particular operating parameter. When the stored value of any operating parameter is displayed, the value

may be changed by pressing the UP/DOWN arrow keys. To select the new parameter value to replace the existing stored value, press the ENTER key. If the operation is successful in replacing the parameter value, the display will read "END" for a short time, and then cycle to the next operating parameter in the selected parameter group.

The GS2 drive has more than a hundred operating parameters that may be programmed, but here we will focus on the most basic to commission a drive. Any operational parameters or keystroke sequences covered here are for example purposes only. These steps should not be considered a substitute for a user's manual, as operating system revisions may require changes. Once you understand the keystroke sequences for programming the VFD you are using, they may be used to complete the practice commissioning problems at the end of this lab.

3. Set motor nameplate voltage
   a. Press the PROGRAM key until the display advances to "P0.00," motor parameters.
   b. This is already the correct parameter selection for this parameter group, so press the ENTER key to display the default value, which is 240 volts.
   c. This parameter may changed to 200, 208, 220, 230, or 240 volts by pressing the ▲ and ▼ keys. After pressing the ENTER key to store the new parameter value, the display should read "END," and then advance to the next parameter selection.
4. Set motor nameplate amperes
   a. Press the PROGRAM key until the display advances to "P0.00," motor parameters.
   b. Press the ▲ and ▼ keys until the display advances to "P0.01," motor nameplate amperes.
   c. Press the ENTER key to display the default value of 2.5 amperes.
   d. This parameter may be set for operation in the range of 0.7 amperes to 2.5 amperes by pressing the ▲ and ▼ keys. After pressing the ENTER key to store the new parameter value, the display should read "END," and then advance to the next parameter selection.
5. Set motor base frequency
   a. Press the PROGRAM key until the display advances to "P0.00," motor parameters.
   b. Press the ▲ and ▼ keys until the display advances to "P0.02," motor base frequency.
   c. Press the ENTER key to display the default value of 60 Hz.
   d. This parameter may be set for 50-, 60-, or 400-Hz operation by pressing the ▲ and ▼ keys. After pressing the ENTER key to store the new parameter value, the display should read "END," and then advance to the next parameter selection.
6. Set motor base RPM
   a. Press the PROGRAM key until the display advances to "P0.00," motor parameters.
   b. Press the ▲ and ▼ keys until the display advances to "P0.03," motor base RPM.
   c. Press the ENTER key to display the default value of 1,750 RPM.
   d. This parameter may be set for operation in the range of 375 RPM to 9,999 RPM by pressing the ▲ and ▼ keys. After pressing the ENTER key to store the new parameter value, the display should read "END," and then advance to the next parameter selection.
7. Set motor maximum RPM
   a. Press the PROGRAM key until the display advances to "P0.00," motor parameters.
   b. Press the ▲ and ▼ keys until the display advances to "P0.04," motor maximum RPM.
   c. Press the ENTER key to display the default value, which is the same as the RPM value stored in memory location P0.03.
   d. This parameter may be set for operation in the range anywhere between the motor base RPM parameter stored in memory location P0.03 and 9,999 RPM by pressing the ▲ and ▼ keys. If no parameter value is set for this memory location, it will default to the parameter value stored in memory location P0.03 (motor base RPM). After pressing the ENTER key to store the new parameter value, the display should read "END," and then advance to the next parameter selection.
8. Set stop methods
   a. Press the PROGRAM key until the display advances to "P1.00," stop methods.
   b. This is already the correct parameter selection for this parameter group, so press the ENTER key to display the default value, which is 00, ramp to stop.
   c. The stop method determines if the drive will stop the motor with a controlled ramp time or if it will allow the motor to coast to a stop. The stored value may be changed with the ▲ and ▼ keys for 00 ramp to stop, or 01 coast to stop. After pressing the ENTER key to store the new parameter value, the display should read "END," and then advance to the next parameter selection.

9. Set the ramp start acceleration time
   The acceleration method of the GS2 is always controlled ramp start, and cannot be changed like the stop method; only the start ramp time may be changed.
   a. Press the PROGRAM key until the display advances to "P1.00," acceleration time 1.
   b. Press the ▲ and ▼ keys until the display advances to "P1.01," acceleration time.
   c. Press the ENTER key to display the default value of this parameter, which is 10 seconds.
   d. The acceleration parameter value may be set anywhere between 0.01 to 600 seconds by pressing the ▲ and ▼ keys. After pressing the ENTER key to store the new parameter value, the display should read "END," and then advance to the next parameter selection.
10. Set the ramp deceleration time
   a. Press the PROGRAM key until the display advances to "P1.00," acceleration time 1.
   b. Press the ▲ and ▼ keys until the display advances to "P1.02," deceleration time 1.
   c. Press the ENTER key to display the default value of this parameter, which is 30 seconds.
   d. The acceleration parameter value may be set anywhere between 0.01 and 600 seconds by pressing the ▲ and ▼ keys. After pressing the ENTER key to store the new parameter value, the display should read "END," and then advance to the next parameter selection.
11. Set the source of operation command
   a. Press the PROGRAM key until the display advances to "P3.00," source of operation command.
   b. This is already the correct parameter selection for this parameter group, so press the ENTER key to display the default value, which is 00, operation determined by the integral keypad.
   c. This parameter determines what the drive will use for control commands. The two possible choices are 00 = operation determined by drive's integral keypad, or 01 = operation determined by external control (start/stop stations, etc.). The stored parameter value may be changed by pressing the ▲ and ▼ keys. After pressing the ENTER key to store the new parameter value, the display should read "END," and then advance to the next parameter selection.
12. Set the external control circuit scheme
   a. Press the PROGRAM key until the display advances to "P3.00," source of operation command.
   b. Press the ▲ and ▼ keys until the display advances to "P3.01," multi-function input terminals.
   c. This memory location is the multi-function input terminal definition for DI, DI2, and sometimes DI3. This parameter value tells the drive what control scheme the drive is wired for. Press ENTER to view the default value of "00," which is basically for two-wire control; "02" is three-wire control for momentary contact pushbuttons. The stored parameter value may be changed by pressing the ▲ and ▼ keys. After pressing the ENTER key to store the new parameter value, the display should read "END," and then advance to the next parameter selection.

   One caveat for this command is that both this setting and the external control circuit wiring of the VFD must match when the drive boots up, or it will go into an external fault and display "EF." When both the setting of this parameter and the external control circuit wiring match, the GS2 must be re-booted so that it recognizes the configuration.

13. Define a fixed operating speed
   There may be an application where the motor only has to run at one fixed speed that is set in advance. This may be accomplished in one of two ways. First, the desired speed may be set with the keypad potentiometer, and then control the stop, start, and reverse functions from the pushbutton station. The problem with setting a fixed operating speed with the keypad potentiometer is that it could be changed by an unauthorized person. A better method is to configure the drive to ignore the keypad potentiometer, and use an operating frequency stored in a memory location of the drive.

   When P3.01 is set to "02," for three-wire control operation, digital inputs DI1, DI2, and DI3 are automatically configured for reading the external pushbuttons for the stop, start, and reversing functions. The next available digital input to use for fixed frequency control, then, is DI4. By default, DI4 is configured for reading the multi-speed memory location P5.01 (where the fixed operating speed frequency will be stored), so no additional configuration of DI4 is necessary to define it as a speed bit.

   a. Connect a control circuit wire between digital inputs DCM and DI4, which will direct the drive to look to P5.01 for the operating speed to run the motor.
   b. Press the PROGRAM key until the display advances to P5.00.
   c. Press the ▲ and ▼ keys until the display advances to P5.01, multi-speed bit 1.

d. Press the ENTER key to display the parameter contents stored at that location, which by default is "0."
e. Press the ▲ or ▼ keys until the display advances to the desired operating frequency.
f. Press the ENTER key to save the new operating frequency value.

14. Define a jog function
There may be an application that requires the motor to operate at its rated 60-Hz frequency, but also have a jog function at a lower speed to position parts of a machine before the process is started. To demonstrate the flexibility of VFDs, this time user-defined digital input #5 will be used for the jog function.
a. Connect an external control circuit for three-wire control, and add a momentary contact jog pushbutton from the DCM terminal, through a momentary contact jog pushbutton, to digital input #5.
b. Press the PROGRAM key until the display advances to P3.00.
c. Press the ▲ and ▼ keys until the display advances to P3.04, multi-function input DI5.
d. Press the ENTER key to display the parameter contents stored at that location, which by default is "04," multi-speed bit 2.
e. Press the ▲ or ▼ keys until the display advances to "09," which is the jog function. After pressing the ENTER key to store the new parameter value, the display should read "END," and then advance to the next parameter selection.
f. The motor jog speed frequency must also be set in operating parameter memory location P5.00. Press the PROGRAM key until the display advances to P5.00.
g. This is already the correct parameter selection for this parameter group, so press the ENTER key to display the default value, which is "6" Hz.
h. Press the ▲ or ▼ keys until the display advances to the desired operating frequency.
i. Press the ENTER key to save the new operating frequency value.

15. Define a DC injection function during startup
There may be an application, such as an air handler where the fan blades may start to turn backward in the airflow when the motor is not energized, when it is necessary to stop a mechanical load before starting the motor. In these situations, DC injection will lock the motor, and therefore the mechanical load, from turning.
a. Press the PROGRAM key until the display advances to P1.00.
b. Press the ▲ and ▼ keys until the display advances to P1.20, the time setting for DC injection during startup.
c. Press the ENTER key to display the parameter contents stored at that location, which by default is "0.0," seconds of applying DC injection.
d. Press the ▲ or ▼ keys until the display advances to the desired number of seconds you want the DC injection to last before the motor is started. After pressing the ENTER key to store the new parameter value, the display should read "END," and then advance to the next parameter selection.
e. The level of current desired for the DC injection must also be set. Press the PROGRAM key until the display advances to P1.00.
f. Press the ▲ and ▼ keys until the display advances to P1.18, DC injection current level.
g. Press the ENTER key to display the parameter contents stored at that location, which by default is "00," percent of DC injection current.
h. Press the ▲ or ▼ keys until the display advances to the desired percentage of DC injection current. After pressing the ENTER key to store the new parameter value, the display should read "END," and then advance to the next parameter selection.

16. Define a DC injection function during stopping
There may be an application that requires the motor to remain locked, and therefore holding the mechanical load locked, for a given time after the motor is stopped. In these situations, DC injection may also be used to lock the motor for a given length of time when the motor is turned off.
a. Press the PROGRAM key until the display advances to P1.00.
b. Press the ▲ and ▼ keys until the display advances to P1.21, the time setting for DC injection during stopping.
c. Press the ENTER key to display the parameter contents stored at that location, which by default is "0.0," seconds of applying DC injection.
d. Press the ▲ or ▼ keys until the display advances to the desired number of seconds you want the DC injection to last after the motor is stopped. After pressing the ENTER key to store the new parameter value, the display should read "END," and then advance to the next parameter selection.
e. The level of current desired for the DC injection must also be set. Press the PROGRAM key until the display advances to P1.00.
f. Press the ▲ and ▼ keys until the display advances to P1.18, DC injection current level.

g. Press the ENTER key to display the parameter contents stored at that location, which by default is "00," percent of DC injection current.
h. Press the ▲ or ▼ keys until the display advances to the desired percentage of DC injection current. After pressing the ENTER key to store the new parameter value, the display should read "END," and then advance to the next parameter selection.
i. DC injection during stopping has one additional operating parameter that must be set, and that is the start-point frequency for applying the DC injection. If the need is to stop the motor as fast as possible and hold it locked, the start-point frequency should be the operating frequency. If the need is to lock the motor for a period of time after it has ramped to a stop, then the start-point frequency should be 0 Hz. Press the PROGRAM key until the display advances to P1.00.
j. Press the ▲ and ▼ keys until the display advances to P1.22, start-point for DC injection.
k. Press the ENTER key to display the parameter contents stored at that location, which by default is "0.0," Hz.
l. Press the ▲ or ▼ keys until the display advances to the desired operating frequency that you want the DC injection current to start. After pressing the ENTER key to store the new parameter value, the display should read "END," and then advance to the next parameter selection.

17. Reverse operation inhibit
There may be an application, such as an impeller deep well pump, that must not turn backward at any time. In these situations it may be desirable to disable the GS2 reversing capabilities.

a. Press the PROGRAM key until the display advances to P6.00.
b. Press the ▲ and ▼ keys until the display advances to P6.03, the reverse inhibit function.
c. Press the ENTER key to display the parameter contents stored at that location, which by default is "00," enable reverse operation.
d. Press the ▲ or ▼ keys until the display advances to the desired operating parameter code, 00 = enable reverse operation, 01 = disable reverse operation. After pressing the ENTER key to store the new parameter value, the display should read "END," and then advance to the next parameter selection.

18. Changing the default display
The GS2 digital display may be programmed to display several different operating parameters when the drive is running a motor load. These operating parameters include 00 = output frequency (Hz), 01 = motor speed (RPM), 03 = output current (A), 04 = motor output current (%), 05 = output voltage (V), and 06 = DC bus voltage (V).
a. Press the PROGRAM key until the display advances to P8.00.
b. This is already the correct parameter selection for this parameter group, so press the ENTER key to display the default value, which is 00, output frequency in Hz.
c. Press the ▲ or ▼ keys until the display advances to the desired display function code. After pressing the ENTER key to store the new parameter value, the display should read "END," and then advance to the next parameter selection. The digital display will now remain on this operating parameter.

19. Resetting the GS2 parameters to factory default values
a. Before programming the operating parameters of the GS2 for this commissioning, reset the drive to the factory default parameter values. Press the PROGRAM key until the display advances to "P9.00."
b. Press the ▲ and ▼ keys until the display advances to "P9.08."
c. Press the ENTER key to view the contents of this parameter.
d. Press the ▲ and ▼ keys until the display reads "99."
e. Press the ENTER key to save this parameter, which will restore the drive to factory default operating parameters. If an external control circuit is wired to the GS2, the display will read "EF" to indicate an external control wiring fault, because the GS2 command control default is the integral keypad, and the drive is still sensing the external control circuit wiring.

20. Connect external control circuit wiring
De-energize the supply power from the GS2, and connect the external control circuit as shown in Figure 16-1. NOTICE THAT THERE IS NO CONNECTION TO ANY TYPE OF POWER FOR THE CONTROL CIRCUIT; DO NOT USE ANY POWER SOURCE FOR YOUR CONNECTION! The GS2 control method senses continuity of the control circuitry between the DCM terminal and each of the digital input terminals for control; it does not require an external power source.

Before starting this exercise, restore the GS2 operating parameters to factory default values, so that no old settings conflict with the new settings.

Figure 16-1 VFD stop/start pushbutton

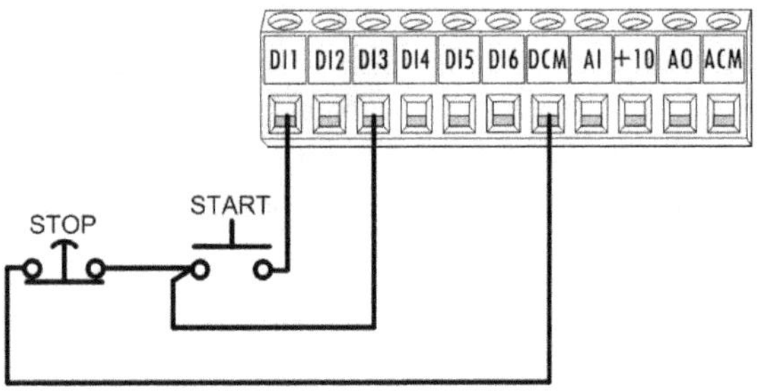

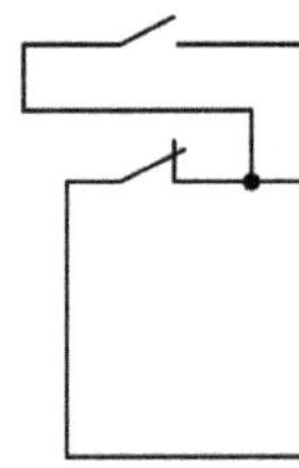

R1O	Relay Output 1 Normally Open
R1C	Relay Output 1 Normally Closed
R1	Relay Output 1 Common
R2O	Relay Output 2 Normally Open
R2C	Relay Output 2 Normally Closed
R2	Relay Output 2 Common
DI1	Digital Input 1
DI2	Digital Input 2
DI3	Digital Input 3
DI4	Digital Input 4
DI5	Digital Input 5
DI6	Digital Input 6
DCM	Digital Common
AI	Analog Input
+10v	Internal Power Supply
AO	Analog Output
ACM	Analog Common

a. After the GS2 drive control circuit is wired as shown in the wiring diagram, apply power to the circuit. If there is an external control circuit wired to the digital input terminals of the drive, it will display a fault code, EF, which means external fault. When the drive was booted, it sensed a wired circuit on the external terminal control wiring terminals that it is not configured for. Whenever wiring changes are made to the external control circuit, the drive must be re-booted so that it can compare its configuration with the external control circuit wiring it senses, and make certain they match.

To complete the GS2 external three-wire control commissioning, there are still two parameters in the P3.00 parameter group that must be modified from default values: operating parameters P3.00, source of operation command, and P3.01, multi-function input terminals for the external control circuit scheme. After the external control circuit is connected, and the two operating parameters in the P3.00 group are changed, re-boot the GS2 to make it operational. Additional operating parameters may need to be changed to optimize the operating parameters for a given application, but the unit is operational.

21. The GS2 relay outputs
The GS2 VFD has two programmable multi-function relay outputs. These relay outputs are rated 120 VAC/24 VDC @ 5 amperes, and 230 VAC @ 2.5 amperes. For the examples in this chapter, 120-VAC indicator lamps will be connected to the relay outputs to demonstrate their function. Note that the programmable multi-function relay outputs are a separate terminal block from the digital input block.

Connect the circuit shown in Figure 16-2, using a green indicator lamp on the normally open contact of relay output #1 and a red indicator lamp on the normally open contact of relay output #2. The other terminals of the indicator lamps get connected together and then connected to the neutral; the ungrounded conductor of the 120 VAC will be applied to the common contact of each relay.

There may be an application where the motor being operated is far away, and out of sight of the operator. The operator, however, has to know that the motor has reached full speed before starting the rest of the process, and the operator also has to know if the motor trips out on overload. The GS2 VFD has two programmable relay outputs for purposes such as this. The green indicator lamp on output #1 will be configured to tell the operator the motor is up to speed, and the red indicator lamp on output #2 will be configured to tell the operator the motor has tripped out on overload.

a. The first configurable multi-function output memory address is P3.11. Press the PROGRAM key until the display reads "P3.00."
b. Press the ▲ and ▼ keys to change to memory location "P3.11" (multi-function output #1).
c. Then press the ENTER key to display the parameter contents. Default is "00" (drive running). Each of the user-defined relay outputs may be programmed to change states by the following operating states: 00 = drive running, 01 = drive fault, 02 = at speed, 03 = zero speed.

Figure 16-2 Control circuit and relay output blocks

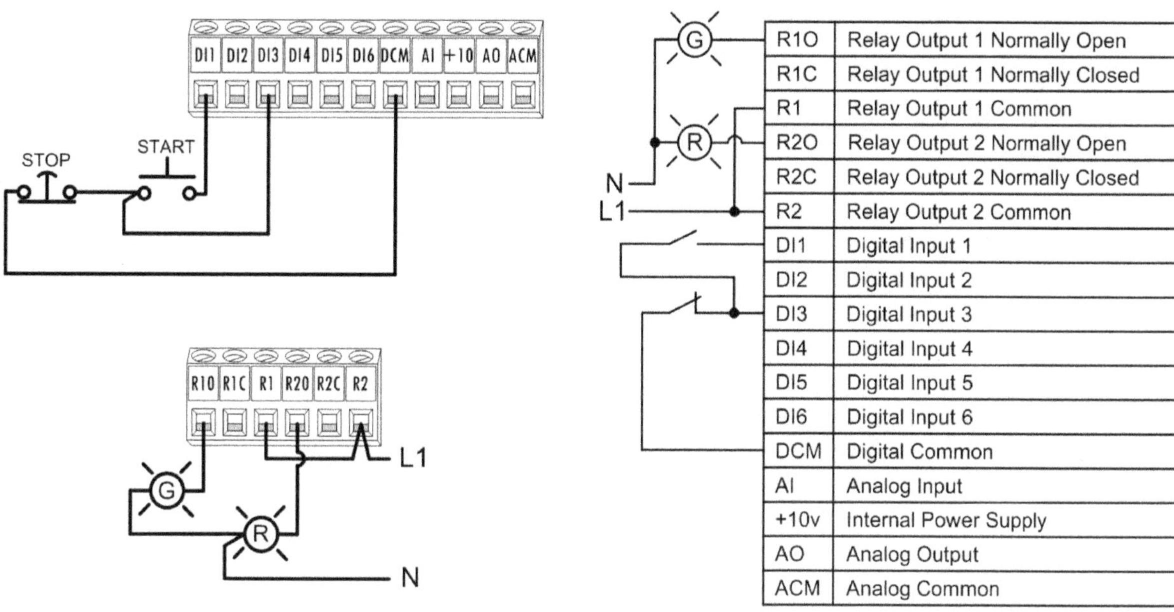

d. Press the ▲ and ▼ keys to change this parameter value to "02" (at speed).
e. Then press the ENTER key to save the new parameter value.

The second configurable multi-function output memory address is P3.12. If you did not return to the display screen after the last step, when you pressed ENTER to save your new parameter value for the first multi-function output, it would have automatically incremented by one to the correct address. If you did not return to the display screen, skip the next two instructions; if you did return to the display screen, continue through the steps as normal.

a. Press the PROGRAM key until the display reads "P3.00."
b. Press the ▲ and ▼ keys to change to memory location "P3.12" (multi-function output #2).
c. Then press the ENTER key to display the parameter contents. Default is "01" (drive at fault). The default value is what we want, so you will not change it.
d. Then press the ENTER key to retain the parameter value.
e. Then press the DISPLAY button to return to the default display.

In order to test the relay output configurations, it is necessary to cause the VFD to go into fault mode because of a motor overcurrent condition. To cause an overcurrent condition you will have to configure the VFD for a lower current than what the motor you are using draws with no load. The memory address to configure the motor current is "P0.01," and the lowest programmable motor current is 0.7 amperes. If you are using a small motor with a very low no-load current draw, it may not be possible to set the motor current low enough.

If the GS2 cannot be set for a lower motor current than the no-load current of the motor you are using, the base frequency, "P0.02," of the GS2 may be set down to 50 Hz. The lower base frequency will decrease the inductive reactance of the motor, and it may cause the motor to draw more current and trip the drive out on overload.

When your configuration is complete, start the motor and confirm that neither lamp lights until the motor reaches full speed, at which time the first indicator light should come on.

Check ☐

Adjust the motor speed to the highest current value possible, which may not be the fastest speed of the motor. Allow the motor to run for several minutes. If the overload condition is small, the VFD will have to run for several minutes before it will cause the VFD to fault on overload.

Check ☐

Notice the display when the VFD goes into the fault condition; it reads "oL I," which means that it was the VFD's internal overload sensing that caused the drive to trip out and stop the motor.

Check ☐

Without resetting the VFD, push the RUN button and try to restart the motor. Note that the motor will not restart until the VFD is reset.

Check ☐

22. Fault records

    The VFD's fault records are stored in memory locations P6.31 thru P6.36, which hold the record of the last six faults the VFD has experienced. To view this record, follow these steps:

    a. Press the PROGRAM key until the display reads "P600."
    b. Press the ▲ and ▼ keys to change to memory location "P6.31" (latest fault).
    c. Then press the ENTER key to display the parameter contents (05 = motor overload, 07 = external fault—these are the only two faults that should be on your VFD).
    d. Press the ENTER key when you are done reading the latest fault. The display will automatically increment by one to "P6.32" (the second most recent fault code).
    e. Press the ENTER key to view this fault code.
    f. Pressing the ENTER key will continue to cycle you through each of the last six faults the VFD has experienced.
    g. Press the DISPLAY key to return to the display mode.

PRACTICE PROBLEMS

Use these practice problems to gain experience configuring the GS2 for different motor application operating parameters. Remember to reset the GS2 to the factory default operating parameters before configuring each of the practice problems to prevent conflicts between saved values from an old configuration and the changed settings of the new configuration.

1. Configure the GS2 for the following motor application operating parameters:
    - Motor nameplate: 60 Hz, 208 volts, 0.9 amperes, 1,760 RPM
    - External control circuit
    - Three-wire stop/start with reverse selector switch
    - Coast to stop
    - Start acceleration: 3 seconds
    - Digital display to read the motor speed in RPM
2. Configure the GS2 for the following motor application operating parameters:
    - Motor nameplate: 50 Hz, 208 volts, 1.4 amperes, 1,740 RPM
    - External control circuit
    - Three-wire stop/start with reverse selector switch, and a jog pushbutton (use the default 6-Hz jog speed)
    - Coast to stop
    - Start acceleration: 2 seconds
    - Digital display to read output current in amperes
3. Configure the GS2 for the following motor application operating parameters:
    - Motor nameplate: 60 Hz, 230 volts, 1.9 amperes, 1,750 RPM
    - External control circuit
    - Three-wire stop/start
    - Ramp to stop
    - Start acceleration: 1 second; stop deceleration: 0.1 second
    - Digital display to read the motor speed in RPM
    - Fixed operating speed of 45 Hz
    - Inhibit reverse operation
    - 5 seconds of DC injection before start (at 100 percent)

4. Configure the GS2 for the following motor application operating parameters:
   - Motor nameplate: 60 Hz, 220 volts, 1.5 amperes, 1,760 RPM
   - External control circuit
   - Three-wire stop/start with reverse selector switch, and a jog pushbutton (set jog speed to 20 Hz)
   - Fixed operating speed of 50 Hz
   - Ramp to stop
   - Start acceleration: 6 seconds; stop deceleration: 6 seconds
   - Digital display of the output current in amperes
   - 5 seconds of DC injection after motor reaches zero Hz (at 100 percent)
5. Configure the GS2 for the following motor application operating parameters:
   - Motor nameplate: 50 Hz, 240 volts, 1.0 amperes, 1,740 RPM
   - External control circuit
   - Three-wire stop/start with reverse selector switch, and a jog pushbutton (set jog speed to 10 Hz)
   - Fixed operating speed of 60 Hz
   - Ramp to stop
   - Start acceleration: 2 seconds, deceleration time
   - Digital display of the motor speed in RPM
   - 3 seconds of DC injection before start (at 100 percent)
   - 15 seconds of stopping DC injection with start-point for DC injection at 60 Hz (at 100 percent)

Correlation of Labs to Text Chapters

Lab Number	Lab Title	Corresponding Text Chapter
1	Analog Meter Usage	None
2	Digital Meter Usage	None
3	Disassemble and Reassemble Motors	2
4	Single- and Three-Phase Power	3
5	Single- and Three-Phase Motor Continuity Tests	5
6	Single-Phase Motor Connections	5
7	Three-Phase Motor Connections	5
8	Motor Nameplates	6
9	Disassemble a Magnetic Relay, and Contactor	7
10	Control Relays	7
11	Magnetic Motor Starters	9, 10
12	Ladder Diagrams	11
13	Two-Wire Control	12
14	Three-Wire Control	13
15	Counting Wires	None
16	Commission an Automation Direct GS2 VFD	16

CPSIA information can be obtained
at www.ICGtesting.com
Printed in the USA
FFHW02n2333070918
48249423-52011FF